T0348467

The Ecology of Poole Harbour

Companion books to this title in the ***Proceedings in Marine Science*** series are:

Volume 1: **Solent Science** – *A Review*
M. Collins and K. Ansell (Eds.)

Volume 2: **Muddy Coast Dynamics and Resource Management**
B.W. Flemming, M.T. Delafontaine and G. Liebezeit (Eds.)

Volume 3: **Coastal and Estuarine Fine Sediment Processes**
W.H. McAnally and A.J. Mehta (Eds.)

Volume 4: **Muddy Coasts of the World: Processes, Deposits and Function**
T. Healy, Y. Wang and J-A Healy (Eds.)

Volume 5: **Fine Sediment Dynamics in the Marine Environment**
J.C. Winterwerp and C. Kranenburg (Eds.)

Volume 6: **Siberian river run-off in the Kara Sea**
Characterisation, quantification, variability and environmental significance
R. Stein, K. Fahl, D.K. Fütterer, E.M. Galimov and O.V. Stepanets (Eds.)

Proceedings in Marine Science **7**

The Ecology of Poole Harbour

Edited by

J. Humphreys
University of Greenwich
UK

and

V. May
Bournemouth University
UK

2005
ELSEVIER
Amsterdam – Boston – Heidelberg – London – New York – Oxford – Paris – San Diego
San Francisco – Singapore – Sydney – Tokyo

ELSEVIER B.V.
Radarweg 29
P.O. Box 211,
1000 AE Amsterdam
The Netherlands

ELSEVIER Inc.
525 B Street, Suite 1900
San Diego,
CA 92101-4495
USA

ELSEVIER Ltd
The Boulevard, Langford Lane
Kidlington,
Oxford OX5 1GB
UK

ELSEVIER Ltd
84 Theobalds Road
London
WC1X 8RR
UK

First edition 2005
Library of Congress Cataloging in Publication Data
A catalog record is available from the Library of Congress.
British Library Cataloguing in Publication Data
A catalogue record is available from the British Library.

ISBN: 0-444-52064-3

♾ The paper used in this publication meets the requirements of ANSI/NISO Z39.48-1992 (Permanence of Paper).

Printed and bound in the United Kingdom
Transferred to Digital Print 2009

Preface

Due mainly to its internationally important bird populations, Poole Harbour has, over recent years, accrued various national and European statutory designations. Whilst these populations provide a relatively conspicuous testament to the harbour as a natural environment, they represent only one aspect of its significance in environmental and ecological terms. The harbour's unique combination of size, hydrological and geomorphological features provide for a rich and productive ecological community. Moreover these same features, in combination with its geographical position on the south coast of England, make it a haven for the naturalization of alien species. In this context, the harbour has been a site for some classic scientific studies, which along with various environmental assessment or monitoring projects and baseline surveys, have added considerably to our knowledge. The need to understand the harbour system, compelling as it is in purely scientific terms, is also necessary as a basis for informed management of the considerable and growing manifestations of human economic activity, and their interactions with this valuable natural resource.

In this context, the editors on behalf of the Poole Harbour Study Group, have sought for the first time to collect together in book form, contributions from various researchers working on the harbour in such a way as to provide, as far as is possible, a picture of the ecology of the harbour system as a whole. As such, this book covers all the major habitats from reedbeds and saltmarshes to the extensive mudflats and the sub-littoral, while also examining in some detail a wide range of ecological phenomena and issues.

Our indispensable starting point for the assembly of the book was a seminar organized by the Poole Harbour Study Group and held on 12 November 2003 at the premises of the Poole Harbour Commissioners. At this seminar, entitled 'The Changing Harbour', nine papers were presented. We are pleased and gratified that all those papers have been written up and now form chapters in this book. In addition to directly providing a substantial part of the book, the seminar had the benefit of generating sufficient interest to enable us to assemble a volume with much more comprehensive coverage than that which could be achieved at a one-day event. Therefore, this book owes its origin above all to those who organized and contributed to that conference and who are appropriately acknowledged on page vi.

The Poole Harbour Study Group aims to enhance our understanding by encouraging and co-ordinating research on Poole Harbour. Our hope is that this volume, by providing a comprehensive, multidisciplinary, whole-system picture of the harbour, will help further these aims.

John Humphreys
Vincent May
March 2005

Contents

Acknowledgements

We would like to thank all those individuals who put in a considerable amount of effort and commitment on both the organization of the seminar 'The Changing Harbour', 12 November 2003, and the subsequent development and publication of this book. These individuals include members of the Poole Harbour Study Group (listed overleaf) but we should acknowledge particularly the work of Maria Pegoraro and the Dorset Environmental Records Centre for taking much of the weight of the organization of the conference along with her colleague, Brian Edwards and Kevin Cook of Fieldwork and Ecological Surveys. Thanks also to Poole Harbour Commissioners for providing space for the conference.

Books such as these are entirely dependent on the willingness of authors to submit contributions. We have been particularly fortunate in getting a group of authors whose aggregate experience and knowledge of Poole Harbour is second to none. We are above all grateful to them.

All but one of the chapters have been written especially for this book. The exception is chapter 6 'History and Ecology of the Cord Grass *Spartina anglica* in Poole Harbour' by Alan Raybould. We are grateful to him and Judy Lindsay, Director of the Dorset County Museum for permission to reprint, in modified form, the article which was first published in the *Proceedings of the Dorset Natural History and Archaeological Society,* Volume 119.

We are grateful to Infoterra, for the satellite photograph used in the Introduction.

Thanks also to the University of Greenwich Marketing Office, in particular Valerie Howe, Peter Birkett and Andrew Beatson for editing, design and production. Also from the University of Greenwich, Avis Brant for a great deal of administrative support in getting the book together.

The contents of a book such as this is dependent on an enormous number of individuals and organizations who have contributed in one way or another to the work that is reported in each of the chapters. Such contributions are acknowledged by individual authors as appropriate at the end of each chapter. We would like to add our own thanks to these individuals and organizations.

John Humphreys
Vincent May

Poole Harbour Study Group

Objectives

The objectives of the Poole Harbour Study Group (PHSG) are to further the study of the geology, hydrology, ecology, physiography and biological communities, and the monitoring of environmental change in Poole Harbour by:

- establishing and maintaining a database of scientific, historical, cultural and other relevant studies of the harbour;
- undertaking, promoting, facilitating, co-ordinating and encouraging further studies of the harbour;
- establishing an archive of harbour studies and keeping this updated;
- facilitating the use of the database and archive by researchers, conservation bodies, statutory bodies, harbour users and others with a bona fide interest in the harbour;
- acting as a centre for advice and information on features of local, national and international interest within the harbour;
- assisting with or sponsoring meetings and publications on ecological, physiographical or biological aspects of the harbour;
- maintaining close links with the Poole Maritime Trust and other bodies with an interest in the harbour environment.

Membership

Sue Burton	English Nature
Richard Caldow	Centre for Ecology and Hydrology
Kevin Cook (Vice Chairman)	Fieldwork and Ecological Surveys
John Day	Royal Society for the Protection of Birds (RSPB)
Anita Diaz	Bournemouth University
Peter Dyrynda	University of Wales
Bryan Edwards	Dorset Environmental Records Centre (DERC)
Neil Gartshore	Royal Society for the Protection of Birds (RSPB)
George Green	Environment Agency
Geoff Hann	National Trust
Tegwyn Harris	Hatherley Laboratories
John Humphreys	University of Greenwich
Antony Jensen	Southampton Oceanography Centre
Vincent May (Chair)	Bournemouth University
Stephen Morrison	Ecological Field Research and Estate Management
Maria Pegoraro	Dorset Environmental Records Centre (DERC)
Angela Peters	National Trust
Bryan Pickess	Royal Society for the Protection of Birds (RSPB)
Eunice Pinn	Joint Nature Conservation Committee
Sally Porter	Poole Harbour Commissioners

Helen Powell	English Nature
Nicole Price	Environment Agency
Martin Slater	Royal Society for the Protection of Birds (RSPB)
Paul St Pierre	Purbeck Biodiversity
Phil Sterling	Environmental Services
Chris Thain	Dorset Wildlife Trust
Peter Tinsley	Dorset Wildlife Trust
Jenny Waldron	Poole
Sarah Welton	Wareham
Suzy Witt	Environment Agency

Poole Harbour Study Group Publications

Cook, K. (2001) *Poole Harbour Reedbed Survey 2000.*

Pickess, B. and Underhill-Day, J. (2002) *The Important Birds of Poole Harbour.*

Edwards, B. (2004) *The Vegetation of Poole Harbour.*

Morrison, S.J. (2004) *Wader and Waterfowl Roost Survey of Poole Harbour: Winter 2002/03.*

Thomas, N.S., Caldow, R.W.G., McGrorty, S., leV dit Durell, S.E.A., West, A.D. and Stillman, R.A. (2004) *Bird Invertebrate Prey Availability in Poole Harbour.*

Chown, D and Cook, K (2004) *Important Breeding Birds of Poole Harbour. Part 1 Water Rail: Part 2 Redshank.*

Correspondence

Dr John Day
Poole Harbour Study Group Secretary
Syldata
Arne
Wareham
Dorset
BH20 5BJ
UK

Website: www.pooleharbourstudygroup.org.uk

Contributors

Sheila Anderson joined the Natural Environment Research Council as a zoologist in 1967, working mainly on coastal and marine issues. She spent 5 years working on intertidal invertebrates in UK estuaries, before joining the Sea Mammal Research Unit to do research on seals. She is now Head of Communications for NERC.

Paola C. Barbuto is an Oceanographer based at Bournemouth University at the School of Conservation Sciences. She is currently conducting an ecological survey of the macro- and mesozooplankton communities in Poole Harbour, with an emphasis on commercial species.

Katie Born is an Environmental Scientist at Halcrow Group Ltd and has worked predominantly on coastal and riverine flood defence schemes for two years. Prior to this, she was an ecologist at the Sheffield Wildlife Trust

Fiona Bowles was the Environmental Manager for Wessex Water, co-ordinating investigations into the environmental impact of the company's operations and new developments. She has worked within the water company's environmental section since 1980 and has now left that role to manage the company's Low Flow Project.

Bronwen Bruce has been working at Dorset Wildlife Trust as their Conservation Officer for Rivers and Wetlands since May 1999. She is responsible for water policy issues, wetland habitat advice and co-ordination of the Dorset Otter Group. With 65 volunteers, this is the largest Otter group in the country.

Richard Caldow works as a Bird Population Ecologist for the Centre for Ecology and Hydrology at Winfrith Technology Centre. His main interest is in investigating the interactions between birds, their food supplies and the environment in which they live in order to predict how bird populations might respond to environmental change.

Ian Carrier has been the Clerk and Chief Fishery Officer of the Southern Sea Fisheries District for 3 years and has worked for the Committee for 13 years. A former Royal Navy Officer, he took the Queen's shilling as an Ordinary Seaman at the age of seventeen before rising through the ranks to Lieutenant Commander prior to 'retiring' in 1991. He has served as a First Lieutenant and Fishery Boarding Officer in the Fishery Protection Squadron and has also spent 3 years in Command of his own ship and squadron. He has a keen interest in all maritime and environmental matters.

Chris Cesar was an Oceanography MSc student in the School of Ocean and Earth Science, University of Southampton. His research project (supervised by Antony Jensen) focused on the impact of clam fishing on the infauna of Poole Harbour

Kevin Cook has had a long interest in Poole Harbour having lived on Brownsea Island for 16 years. He is Chairman of the Dorset Environmental Records Centre and Vice Chairman of the Poole Harbour Study Group. He works as a freelance Ecologist for his own company Fieldwork Ecological Services Ltd.

Anita Diaz is a Senior Lecturer in Ecology at Bournemouth University. She has research interests in conservation biology and has a number of ongoing research projects based around Poole Harbour. These include investigation of the ecological impact of Sika Deer and examination of methods for heathland ecosystem re-creation on agricultural land.

Sarah E. A. Le V. dit Durell is a Wader Population Ecologist at the Centre for Ecology and Hydrology Dorset, studying shorebirds and their invertebrate prey. She has been involved in several surveys of macrobenthic invertebrates in Poole Harbour. Her main contribution to these surveys is invertebrate identification and measurement.

Peter Dyrynda is a Marine Biologist based at University of Wales Swansea. He has a long association with the Dorset coast, and more specifically with Poole Harbour and the Fleet Lagoon. He has undertaken many of the sub-tidal appraisal and monitoring studies undertaken in Poole Harbour over the past two decades.

Bryan Edwards is the Senior Surveyor with the Dorset Environmental Records Centre where he has worked since 1991 carrying out habitat and species surveys, both in Dorset and outside the county. His main interests are botany, particularly lichens and bryophytes, and plant communities. He is co-author of *The Mosses and Liverworts of Dorset.*

Paul English is a Senior Marine Ecologist and the manger of the marine biological laboratory at Emu Ltd, Hampshire. He is a specialist in the study of marine macrofaunal sediment communities and impact prediction.

Andy Gale is Professor of Geology at the University of Greenwich, and has a lifelong interest in the geology of southern England, specializing in the stratigraphy and palaeontology of Cretaceous and Tertiary sediments.

Justine Hannaford is a recent graduate from the BSc (Hons) Environmental Protection course at Bournemouth University. She carried out some of the work presented in her chapter as part of her third year project under the supervision of Anita Diaz.

Craig House is a recent graduate from the BSc (Hons) Environmental Protection course at Bournemouth University. He carried out the work presented in his chapter as part of his third year project under the supervision of Anita Diaz and Vincent May.

John Humphreys is Pro Vice-Chancellor and Professor at the University of Greenwich. His scientific research focus is on estuarine ecology with an emphasis on invertebrate benthic populations.

Antony Jensen is a senior lecturer in the School of Ocean and Earth Science at the University of Southampton. His current research interests include artificial reefs, inshore fisheries and benthic ecology. He has served on the Southern Sea Fisheries Committee as a DEFRA appointee since 1997 and is currently Vice Chairman of the Eastern Sub-committee.

Martin Jones returned to higher education after 20 years in engineering. As a lifelong resident of Poole and a regular user of the harbour, he hopes that the algal mat project will play a part in conserving the fragile ecological balance of one of the UK's finest natural harbours.

Selwyn McGrorty has 35 years experience of research into the ecology and population dynamics of coastal, intertidal invertebrates around Britain and northern Europe. He is currently working at the Centre for Ecology and Hydrology at Winfrith, Dorset on predictive models of shorebirds and their prey, particularly shellfish.

Vincent May is Emeritus Professor of Coastal Geomorphology and Conservation at Bournemouth University, and has been investigating the harbour since the mid-1960s. He chairs the Poole Harbour Study Group and the Purbeck Heritage Committee.

Linda C. Parker is an independent consultant in benthic ecology. She continues to work on the effects of pump-scoop dredging on the intertidal communities at Poole Harbour.

Bryan Pickess is a Field Naturalist, who for over 30 years managed the RSPB's Arne Nature Reserve, which borders Poole Harbour. He retired in 1996 and continues to be involved with the ecology and management of lowland heath, the birds of the harbour and other wildlife conservation management matters. He is the co-author of *The Important Birds of Poole Harbour.*

Eunice H. Pinn is a Fisheries Advisor for the Joint Nature Conservation Committee. Prior to this she was a Senior Lecturer in Coastal Zone Management at Bournemouth University. Her research has focused on factors affecting marine biodiversity including the effect of anthropogenic activities such as fishing in the marine environment and the impact of faunal bioturbation and boring activity on biodiversity and biogeochemical cycling.

Helen Powell is a Conservation Officer at English Nature's Dorset Team. She is responsible for Poole Harbour Site of Special Scientific Interest and other coastal and freshwater SSSIs in south east Dorset.

Alan Raybould studied the genetics and ecology of *Spartina anglica* for 16 years while working at the Institute of Terrestrial Ecology's Furzebrook Research Station (now part of the Centre for Ecology and Hydrology Dorset). He now works on environmental safety assessments of genetically modified crops for Syngenta.

Neil Richardson joined the team of the Southern Sea Fisheries District in 2001. Most of his life has been connected with the sea. Together with his father, he was joint owner of a commercial fishing vessel for 6 years before selling up to go to university where he studied coastal zone management. During this time, he carried out a work placement aboard the Fishery Protection vessel of the South Wales Sea Fisheries Committee and on graduating in 1996, he commenced work as a Fishery Officer with them.

Richard Stillman works as a Bird Population Ecologist and Modeller for the Centre for Ecology and Hydrology Dorset. His main interest is to predict how bird populations are influenced by environmental change from an understanding of the foraging behaviour of the individual animals within these populations.

John Underhill-Day has worked as an Ecologist and Land Manager for the Royal Society for the Protection of Birds for 35 years. He has lived and worked on the edge of the harbour for 9 years and has been the Secretary of the Poole Harbour Study Group since its formation. He is co-author of *The Important Birds of Poole Harbour*.

Julian Wardlaw has been responsible for pollution prevention and response, enforcing water pollution legislation and emergency planning within and around Poole Harbour since 1980, currently with the Environment Agency.

Andrew West is an Ecologist at NERC Centre for Ecology and Hydrology Dorset. He works mainly on computer models of shorebird populations and has also taken part in intertidal surveys of a number of British estuaries.

The Ecology of Poole Harbour
John Humphreys and Vincent May (editors)

Introduction: Poole Harbour in Context

John Humphreys[1] and Vincent May[2]

[1]University of Greenwich, Old Royal Naval College, Greenwich, London SE10 9LS

[2]School of Conservation Sciences, Bournemouth University, Talbot Campus, Fern Barrow, Poole, Dorset BH12 5BB

Throughout the world, coastal ecosystems are at risk. Over half of the world's coastlines suffer from severe development pressure predicated on the growth of human populations and the increasing propensity of those populations to congregate in coastal areas. With coastal urbanization has come rapid industrial and commercial development which has put increasing pressure on coastal wetland habitats including estuaries, mudflats and saltmarshes (World Resources Institute, 1995).

Although accurate figures do not exist, it is possible that half the earth's natural coastal wetlands have gone. In any event it is known that since the Second World War many millions of hectares have succumbed to urban expansion, land reclamation and drainage for agriculture among other things (Hinrichsen, 1998). Furthermore, many of those coastal ecosystems that remain suffer reduced vitality and viability as a consequence of pressures derived directly from human activity, for example, serving as ports, recreation centres, fishing grounds or receivers of effluent. Yet such coastal environments are ecologically and economically significant assets. Estuaries and saltmarshes are important in terms both of biodiversity and as highly productive natural ecosystems whose significance ranges well beyond their immediate locality. They are spawning and nursery areas for commercially harvested species of fish and provide essential seasonal feeding grounds for bird populations whose migrations range across continents. As such it is important that coastal ecosystems are well understood and sustainably managed.

All of the above generalizations apply in the particular case of Poole Harbour (Figure 1). Indeed Poole Harbour is arguably unusual in the extent to which it represents in microcosm coastal zone issues in the developed world. The contrasting and conflicting pressures on Poole Harbour are sometimes startling. For instance, the harbour entrance – no more than 370 m wide – separates on the south-west side an unspoilt protected natural environment of considerable importance (Studland) from, on the north-east side, a residential centre where property competes with Manhattan and Hong Kong island in the world-wide table of real estate values (Sandbanks), see Figures 1 and 2 (Concoran, 2000).

Figure 1 **Satellite photograph of Poole Harbour and surrounding areas. Note the contrast of urban development to the north and west with natural and rural environments to the south and east. (Image courtesy of Infoterra.)**

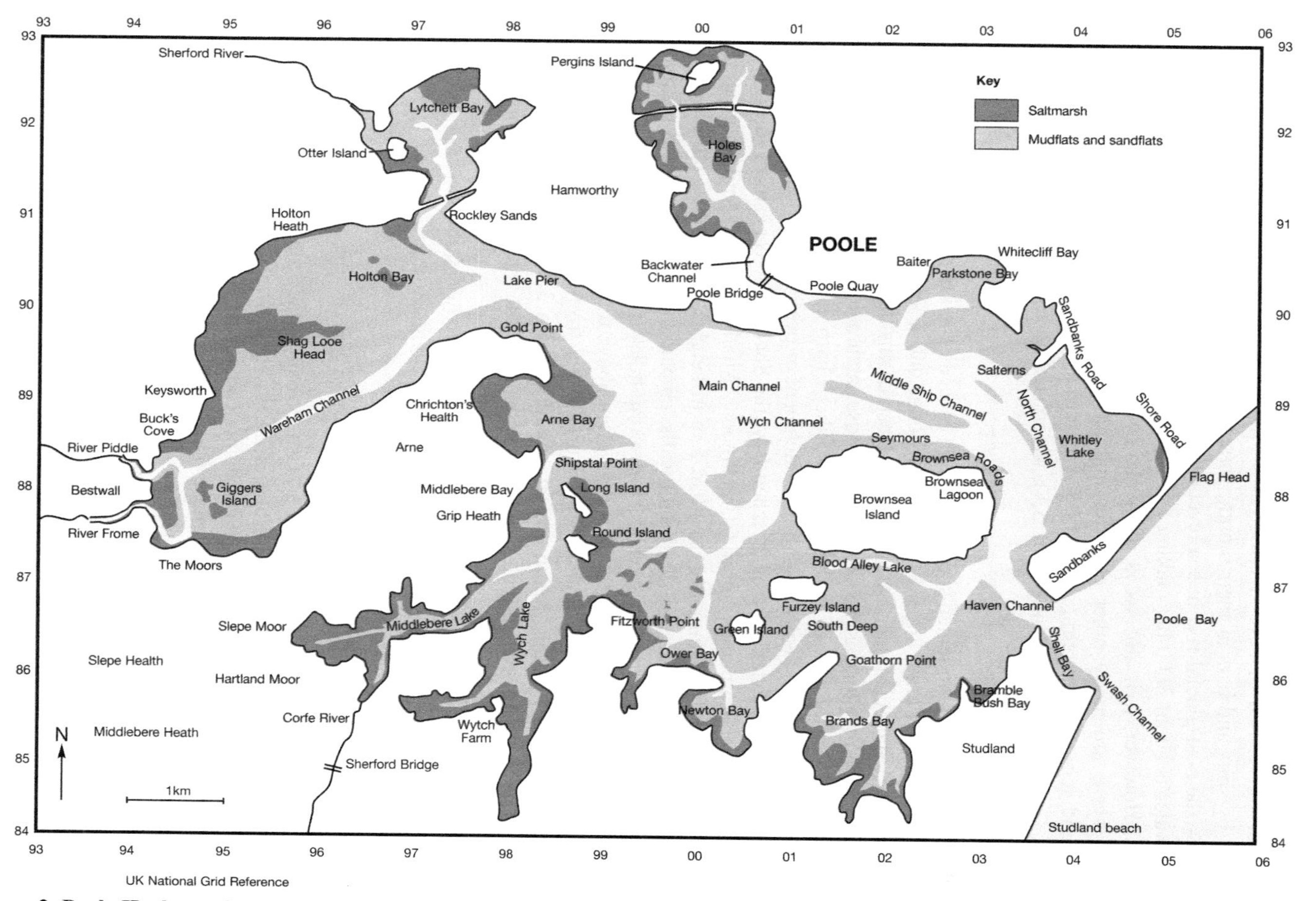

Figure 2 Poole Harbour showing the approximate locations of the main features referred to in this book.

General description

Poole Harbour is one of several estuaries on the south coast of England which are enclosed by spits and bars at their mouth, formed as a result of the drowning of river valleys by the post-glacial rise in sea level (Figure 3). At High Water Spring Tides, the area of water is about 3600 ha, making Poole Harbour one of Europe's largest lowland estuaries. However, it is not, as is often claimed, the world's second largest natural harbour. There are many other estuaries and lagoons which have narrow restricted mouths and serve as harbours, although they may not be so-named. They include, for example, Grays Harbor (Washington, USA), which is almost six times the area of Poole Harbour.

The harbour has a long indented shoreline which exceeds 100 km and there are five main islands (Brownsea, Furzy, Green, Round and Long) and a single entrance (Figure 2). There are three main channel networks. South Deep drains the southern lowland heaths, and the Wych Channel drains the Corfe River whose catchment is mainly on the Wealden sands and clays of the Isle of Purbeck. The northern harbour forms the estuary of the Rivers Frome and Piddle (with a combined catchment of over 770 km^2) and two smaller embayments with restricted mouths, Lytchett Bay (the Sherford River estuary) and Holes Bay (draining the heathlands around Creekmoor). Much of the natural shoreline is marked by a low bluff (commonly less than 5 m in height) and eroding cliffs, but the northern shoreline is mainly artificial with walls, embankments, marinas and wharves. The harbour has a small tidal range (1.8 m at spring tides, 0.6 m at neaps) and a double high water which means that water levels are often above mean tide level for 16 out of 24 hours. Mean monthly maximum temperatures range from 8 °C (January) to 27 °C (August) and mean monthly minimum temperatures from 3 °C (February) to 16 °C (August). Ground temperatures can fall below freezing on the intertidal flats and rise to over 30 °C in summer, but there have been few local studies of the estuary's microclimate. Although winds are mainly from the west or south-west, they can be modified in summer by sea breezes from the south or south-east.

History of human activity

Poole Harbour has been used for trade and fisheries since the Iron Age, and its history reflects its role as a port and the exploitation of the natural resources both within the harbour and around its shores. Late Iron Age and Romano–British pottery, salt-working, iron smelting and shale working took place on the southern shore and on Green Island. Imported goods (pottery and amphorae) found here and the structures of a late Iron Age port make Poole one of the earliest cross-channel trading ports (Markey *et al.*, 2002). Upstream there was extensive medieval reclamation of the Frome and Piddle floodplains and the upper estuary.

By the thirteenth century, Poole was a prosperous commercial port. From the late seventeenth century to the early nineteenth century, Poole thrived on its Newfoundland

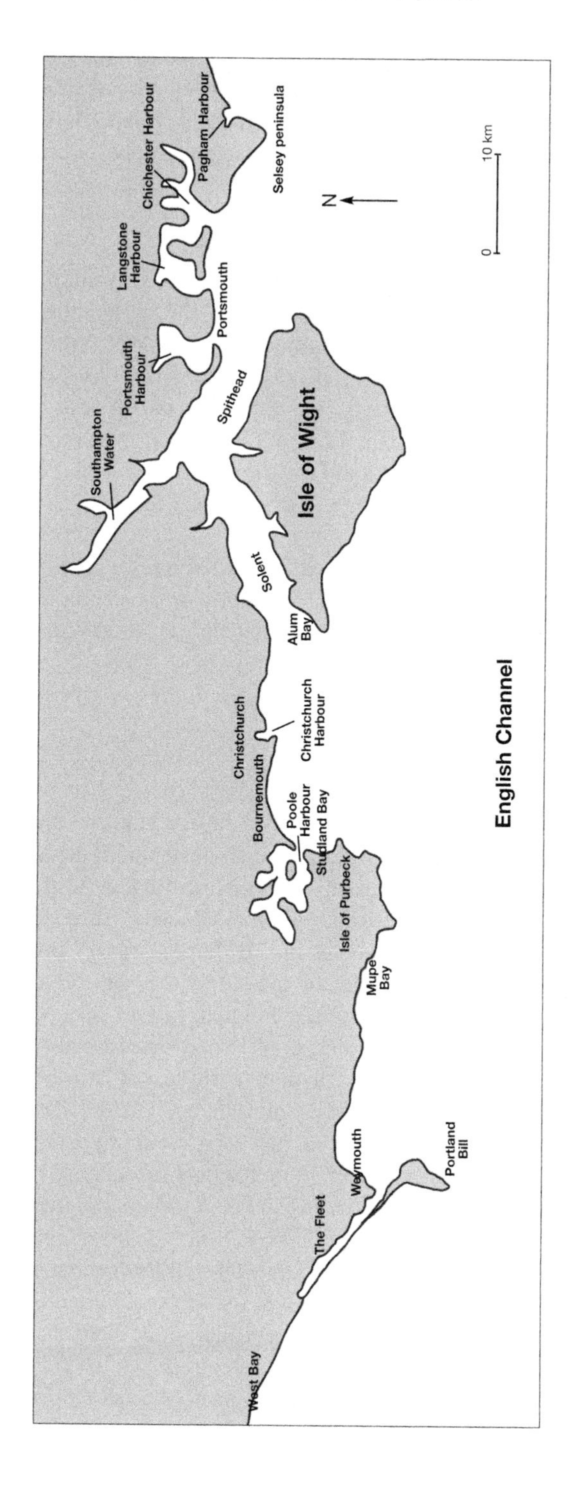

Figure 3 Map showing the position of Poole Harbour in relation to the adjacent coast of southern England.

trade and reached the peak of its prosperity. In 1802, the Newfoundland trade employed over 350 ships with 2000 men and there were at least twelve boatyards. However, with Napoleon's defeat, blockades ended, and Poole's fish trade collapsed. Although demand for clay for potteries and the coastal grain trade expanded, the port declined. The Poole oyster fishery also collapsed.

The Second World War saw the harbour used intensively. International flying boat services (both by Imperial Airways and the RAF) were moved from Southampton. The harbour was an important base for preparations for the D-Day landings in 1944. From the mid-1950s, development of the shore continued with a power station at Hamworthy (now demolished), reclamation and waterfront building for recreation, roads and marinas. There are currently eight yacht clubs and ten boatyards as well as marinas associated with residential developments. Europe's largest onshore oilfield lies beneath the harbour with wells on Furzy Island and Goathorn and the port has been enlarged to provide for roll-on-roll-off freight and larger cross-channel ferries.

Throughout its history, the harbour has been recognized as a resource to be exploited and altered to enhance the prosperity of the town. Any study of the ecological history of the harbour thus needs to acknowledge the human impacts over 2000 years on the harbour ecosystem.

Ecological character

Poole Harbour contains a wide variety of habitats not least due to its size, estuarine gradients and the fact that most of the harbour bed falls within the intertidal zone. As Gray (1985) observed, the harbour and its adjacent shores contain most types of British coastal habitat within an area which is relatively small in relation to this diversity. The intertidal area includes extensive mud and sandflats mostly fringed by reedbeds or saltmarshes (Figure 2). Much of the saltmarsh area is dominated by Cord Grass *Spartina anglica* and the harbour has been the site of classic studies on the development and decline of this species since its first occurrence. Eelgrass *(Zostra* sp.) beds are known to occur in the sub-littoral. Cobbles, stones and gravel occur in channels subject to scour, and sub-tidal bedrock occurs in the Haven Channel to the east of Brownsea Island. Much additional hard substrate, however, has been introduced into the harbour by human activity. This includes boulders introduced as part of a causeway to the south-west of Green Island and increasingly concrete and stone habitats associated with quay, marina and other construction, especially along the urban north-west perimeter.

Since 1999, Poole Harbour has been classified as a Special Protection Area (SPA) under the European Union Birds Directive on the basis of its internationally important population of birds. The harbour and its adjacent landscape hold a number of other European and national statutory designations which serve to protect the natural environment, including that of a designated European Marine Site. Despite its high conservation value, the harbour has a history of problems of contamination and hypernutrification, which appears to have affected species abundance and distribution while also reducing biodiversity in some habitats (Langstone *et al.*, 2003).

References

Concoran (2000) *International Report for 2000*. New York: Concoran.

Gray, A. J. (1985) *Poole Harbour: Ecological Sensitivity Analysis of the Shoreline*. Monks Wood: Institute of Terrestrial Ecology.

Hinrichsen, D. (1998) *Coastal Waters of the World*. Washington DC: Island Press.

Langstone, W. J., Chesman, B. S., Burt, G. R., Hawkins, S. J., Readman, J. and Worsfield, P. (2003) *Characterisation of European Marine Site: Poole Harbour Special Protection Area. Occasional Publication* No 12. Plymouth: Marine Biological Association of the UK.

Markey, M., Wilkes, E. and Darvill, T. (2002) Poole Harbour. Iron Age port. *Current Archaeology*, **181**: 7.

World Resources Institute (1995) *WRI Indicator Brief: Coastlines at Risk: An Index of Potential Development and Related Threats to Coastal Ecosystems*. Washington DC: World Resources Institute.

1. The Geology of Poole Harbour

Andy Gale

Department of Earth and Environmental Sciences, School of Science, University of Greenwich at Medway, Chatham Maritime, Kent ME4 4AW

e-mail: Ga14@gre.ac.uk

The last 400 million years of the geological history of Poole Harbour and the region adjacent to it are reviewed. This is facilitated by the extensive exploration for oil which has been undertaken by BP in the development of the Wytch Farm oilfield, and the exploration wells made by English China Clay in search of economically viable deposits of kaolinite. The closure in the Carboniferous of the Rheic Ocean lead to the formation of east-west thrusts, now deep beneath southern England, which have determined the structural grain of the region. A series of en echelon thrusts run beneath east Dorset and the Isle of Wight. These thrusts have been reactivated extensionally during the opening of the Atlantic (Jurassic–Cretaceous), and compressionally during the early Alpine Orogeny in the Palaeogene, when collision between Africa and Europe commenced. Hydrocarbon accumulations were formed by burial of organic-rich Early Jurassic sediments on the southern, downthrown, side of a fault. This burial resulted in maturation of oil, which subsequently migrated up the fault plane to become trapped beneath Wytch Farm. Inversion in the Palaeogene, 40–50 million years ago, created the Hampshire Basin, and caused the uplift of the Chalk ridges of Purbeck and the Isle of Wight. A Proto-Solent drainage pattern, with a catchment to the north and west, thus developed in the Palaeogene. This brought kaolinite from the granites of Devon into the region, and valuable 'ball clays' were deposited during the Eocene Period. The Holocene transgression, 10,000 years ago, caused the sea to break through the Chalk ridge joining the Isle of Wight and Dorset, and created the western Solent Channel.

Introduction

From the viewpoint of a casual observer, the geology of Poole Harbour appears to be less than spectacular. The low topography of the region with few poor exposures of the underlying low-dipping, rather unfossiliferous Eocene sands and clays (Figures 1 and 2), river terrace gravels of Pleistocene age and extensive Holocene alluvium (see the BGS *Bournemouth Solid and Drift Geology Sheet* No. 339 and the *Bourmemouth Sheet Memoir*; Bristow *et al.*, 1991) does not attract many geological visitors. Poole Harbour and its surrounding countryside and offshore region, however, have the most intensively studied geology in the entire south of England. This has been driven by two separate economic geological imperatives. Firstly, and more importantly, the discovery in 1973 of the Wytch Farm oilfield in the Jurassic Bridport Sands, 1500 m beneath Poole Harbour, encouraged extensive exploration of the region, with the use of deep boreholes and seismic profiles and lead subsequently to major discoveries in the Triassic Sherwood Sandstone (Underhill and Stoneley, 1998). Wytch Farm is the largest onshore oilfield in Europe, with estimated reserves of 300 million barrels (Buchanan, 1998). Much of the geological data obtained during exploration have now been published and enable a detailed reconstruction of the last 250 million years of geological history of the region, which is a microcosm of the history of southern England and the English Channel. Additionally, Permian, Triassic and Cretaceous rocks, essentially identical to those which underlie Wytch Farm, are extensively exposed in the sea cliffs of east Devon and Dorset (House, 1993), and thus allow detailed study at outcrop on what has become recently a UNESCO World Heritage Site.

The presence of layers of valuable clays in the Eocene Poole Formation in the area of Wareham and Poole Harbour (so-called 'ball clays') has resulted in extensive shallow exploration by English China Clays, and numerous open cast workings for these clays exist at the present time, as on the Arne peninsula. This exploration has resulted in a detailed stratigraphical understanding of the Eocene sediments beneath Poole Harbour (Bristow *et al.*, 1991).

This chapter aims to provide the broad geological context and history, not just of Poole Harbour and its immediate environs, but also of the broader region of Wessex, including Dorset, south Hampshire and the Isle of Wight (Figure 1). I will try to show how the

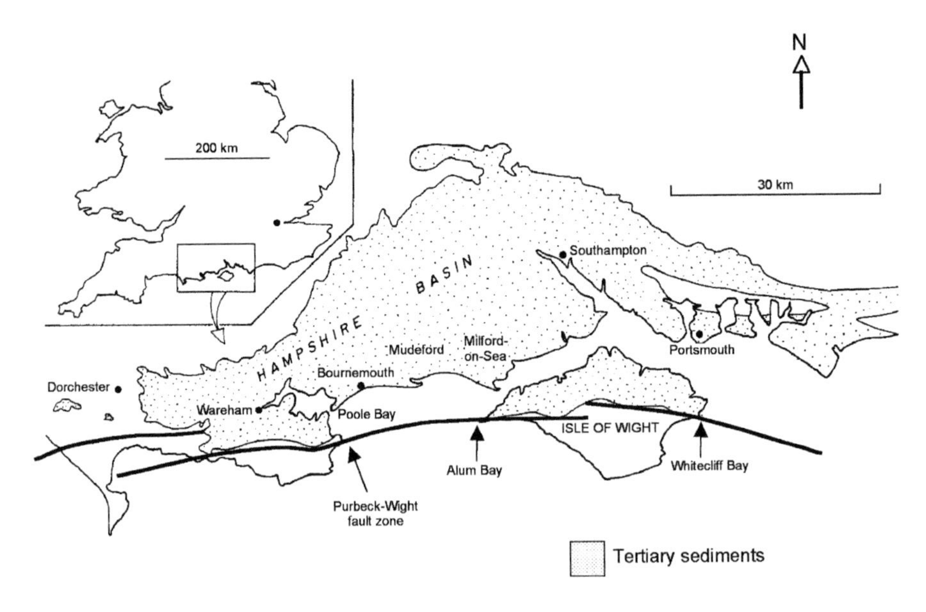

Figure 1 Geological and structural map of southern central England, to show the Hampshire Basin and the axes of anticlines which extend east-west through the Isle of Wight and Isle of Purbeck.

geological evolution of the region has lead to the sedimentary succession found in the region, the local landforms and geomorphology, and the formation and preservation of hydrocarbon. reserves.

Geological history of the Poole Harbour region

Closure of the Rheic Ocean

At 2700 m beneath Poole Harbour lie the oldest rocks known in Dorset – Devonian phyllites (silky-sheened recrystallized slates) found at the base of the deepest Wytch Farm wells (Figure 3). Closure of the Mid-European or Rheic Ocean in the Permian led to the Variscan Orogeny or period of mountain building when the phyllites formed through the processes of low-grade metamorphism (Warr, 2000). An important consequence of the Variscan Orogeny was the formation of east-west thrust faults, which now on the Variscan Front lie in the Palaeozoic Basement deep beneath southern England and northern France. These south dipping structures cut hard, brittle Devonian and Carboniferous limestones, and are well exposed in the quarries in the Palaeozoic inlier in the Boulonnais in the Pas de Calais, northern France. The ancient thrusts have acted as planes of weakness which have been reactivated again and again during the periods of extensional and compressional stress affecting the region. The prominent east-west ridges of the Purbeck Hill, Abbotsbury and the Isle of Wight seen on present day satellite images are a consequence of this deep structural grain.

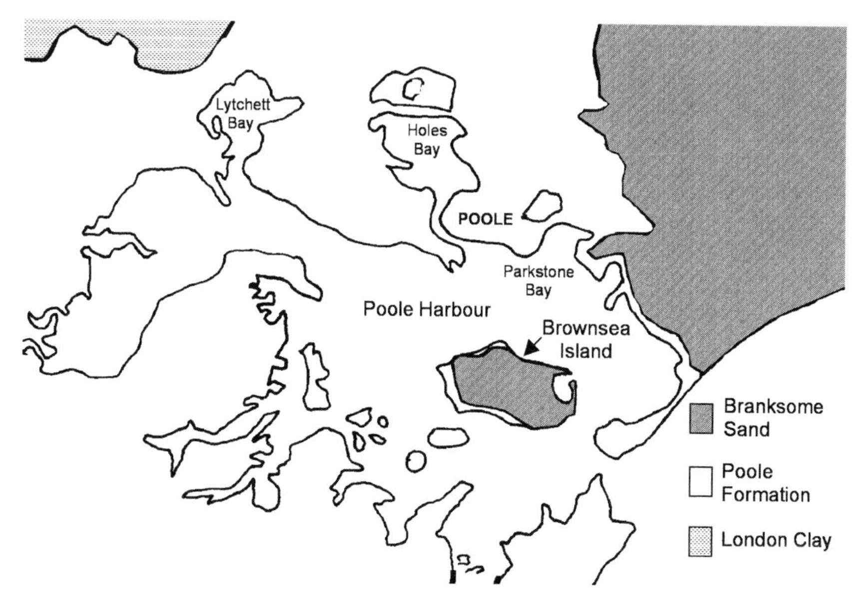

Figure 2 Geological map of Poole Harbour to show the distribution of solid deposits of Eocene age.

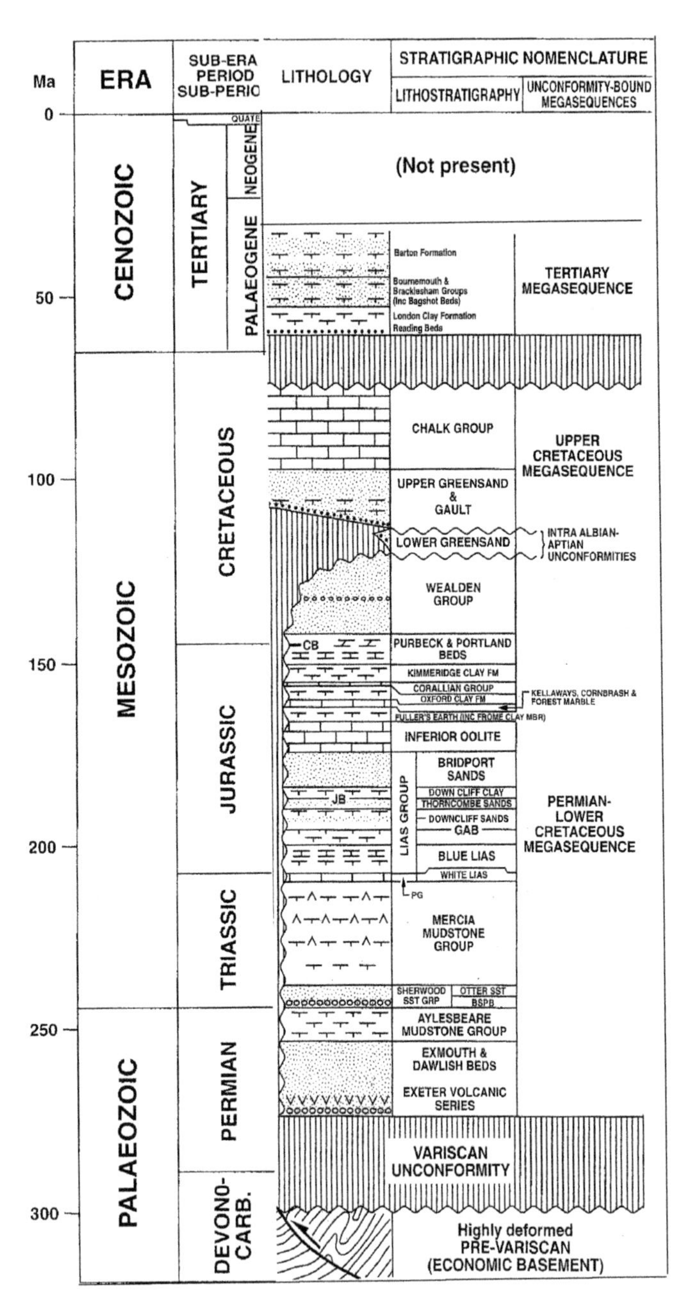

Figure 3 Geological history seen in the Wytch Farm wells. Note that after the substantial Variscan Unconformity, the succession of Triassic and Jurassic rocks is substantially complete. Note the unconformities of Early Cretaceous and Late Cretaceous-Palaeogene age. The Sherwood Sandstone and Bridport Sands form the reservoirs for oil at Wytch Farm. From Underhill and Stoneley (1999) with permission.

Triassic rifting and the initial opening of the Atlantic

Around 300 million years ago, in the Late Carboniferous, Europe and North America were part of the supercontinent Pangea. Shortly later, in the Early Permian, initial rifting of the North Atlantic commenced in the vicinity of the present Rockall Trough (Ruffell and Shelton, 2000). At the same time, the Tethys Ocean was opening in the south of Europe between Spain and Africa. The net consequence of this extensional pull apart was the formation of rift basins in which considerable quantities of Permian and Triassic sediment accumulated. In the Wessex Basin, beneath Dorset, Hampshire and the Isle of Wight, the deep Variscan structures became normal faults, and the Channel area was the focus of thick non-marine sedimentation through the latest Permian and Triassic. Approximately 1500 m of conglomerates, mudstones and sandstones deposited by flash floods, braided rivers and playa lakes are found in the Wytch Farm wells. The fluvial Sherwood Sandstone is the most intensively studied of these, for its higher porosity meant that it was to become the major reservoir for oil much later on. It lies between the Aylesbeare and Mercia Mudstones, and the latter would subsequently provide an impermeable cap, trapping the Wytch Farm oil. These deposits are magnificently exposed on the south Devon coast between Exmouth and Seaton, and the section to the east of Sidmouth in particular shows how successive braided river channels in the Sherwood Sandstone were cut, filled and abandoned.

Opening of the central part of the Atlantic, with generation of oceanic basaltic crust, was roughly coincident with the base of the Jurassic, 200 million years ago. The result of this event was increased stretching of the European continental margin with resultant thermal subsidence. A major marine transgression ensued, and the Early Jurassic deposits are the dark marine shales of the Lias which covered most of southern England, excepting the London Platform (Hesselbo, 2000). The Blue Lias (named after the quarrymen's dialect 'lias' meaning layers) comprises decimetre-scale alternations of dark, laminated organic-rich shale and limestone, essentially identical to the facies developed in the Wytch Farm wells. The Blue Lias is exposed in the crumbling cliffs to the west of Lyme Regis, and is famous as the source of superbly preserved fossil reptiles seen in the fossil shops of Lyme Regis and in various museums throughout the country. The Lias is the source rock for all the oil found in the Wessex Basin, including Wytch Farm, and coincidentally, the factors leading to the preservation of both the entire reptile skeletons and the oil were basically the same. The Liassic Sea periodically became stagnant, and the bottom waters almost entirely depleted in oxygen. As a result, the aerobic bacteria which destroy organic matter were unable to exist, and animal carcasses were not scavenged, nor was much organic matter broken down by anaerobic bacterial action.

The Jurassic succession in Wytch Farm is similar to that exposed on the Dorset coast, and it comprises clays, sands and thin limestones which can be examined at outcrop between Lyme Regis and Weymouth (House, 1993). The Bridport Sand in particular has received special attention in view of its importance as a reservoir in Wytch Farm. On the Dorset coast, to the east of West Bay, the Bridport Sand forms striking yellow-brown vertical sandstone cliffs, capped by the thin Inferior Oolite. The rather soft highly burrowed marine sands contain hard concretionary layers every metre or so, and these weather out to form ledges. The Bridport Sand is almost identical beneath Poole Harbour, where the hard concretions may pose a problem for reservoir engineers because they restrict the vertical flow of oil. The highest Jurassic formation in Wytch Farm is the Oxford Clay, which is truncated by an erosional unconformity (Figure 3).

Formation of the Channel Basin

Renewed extension during the Late Jurassic resulted in continued reactivation of the deep Purbeck–Isle of Wight lineaments as normal faults, and formation of the Channel Basin to the south of the faults. This basin received continuous sedimentation during the Early Cretaceous, and a thick non-marine (essentially fluvatile) Wealden and marine Lower Greensand succession was deposited in the south of the Isle of Wight and in south-east Dorset. On the hanging wall of the fault, Jurassic rocks were actively eroding and supplying sediment to the basin fill. Close to the fault in the eastern Isle of Wight, Jurassic fossils are found derived into Early Cretaceous sediments (Radley *et al.*, 1998). Hanging wall uplift and erosion continued in the Poole Harbour area, such that the highest Jurassic was stripped off down to the Oxford Clay. Movement on the faults ceased in the Early Albian, and the areas north of the fault were transgressed by a shallow sea in which the Lower Greensand and Gault Clay Formations were deposited. Thus, the Lower Greensand rests unconformably upon Oxford Clay in the Poole Harbour region (Figure 3).

The Late Jurassic–Early Cretaceous normal movement on the Purbeck and Isle of Wight faults (and probably others, including the Abbotsbury fault), had a further consequence; on the southern footwalls of these faults, the organic-rich shales of the Lias source rock was taken down into the 'oil kitchen', the depth at which temperatures were sufficient to generate petroleum (Selley and Stoneley, 1987; Underhill and Stoneley, 1998). The newly formed oil migrated upwards during the Cretaceous, and spectacular 'fossil' oil seeps can be seen on the Dorset coast, east of Osmington Mills and at Mupe Bay, east of Lulworth Cove. Here, sands are cemented by sticky dark brown residual oil which escaped up faults, probably during the Cretaceous. It appears that oil actually cemented the banks of an Early Cretaceous river at Mupe Bay. Only at Wytch Farm did large quantities of oil migrate across two faults to become trapped in the tilted reservoirs of the Sherwood Sandstone and the Bridport Sands (Figure 4). Elsewhere, the bulk of the hydrocarbons was lost by leakage through the fault systems. The Kimmeridge Bay oilfield is the only example known in the region where migration occurred later on, in the Tertiary.

The base of the Late Cretaceous, 100 million years ago saw the commencement of a major marine transgression which was caused by a combination of factors (Gale, 2000a,b). Firstly, the rate of sea floor spreading increased globally and more water was displaced from the ocean basins by newly formed mid-ocean ridges. Secondly, tectonic subsidence of the Atlantic margin lead to increased marine transgression. Thirdly, the

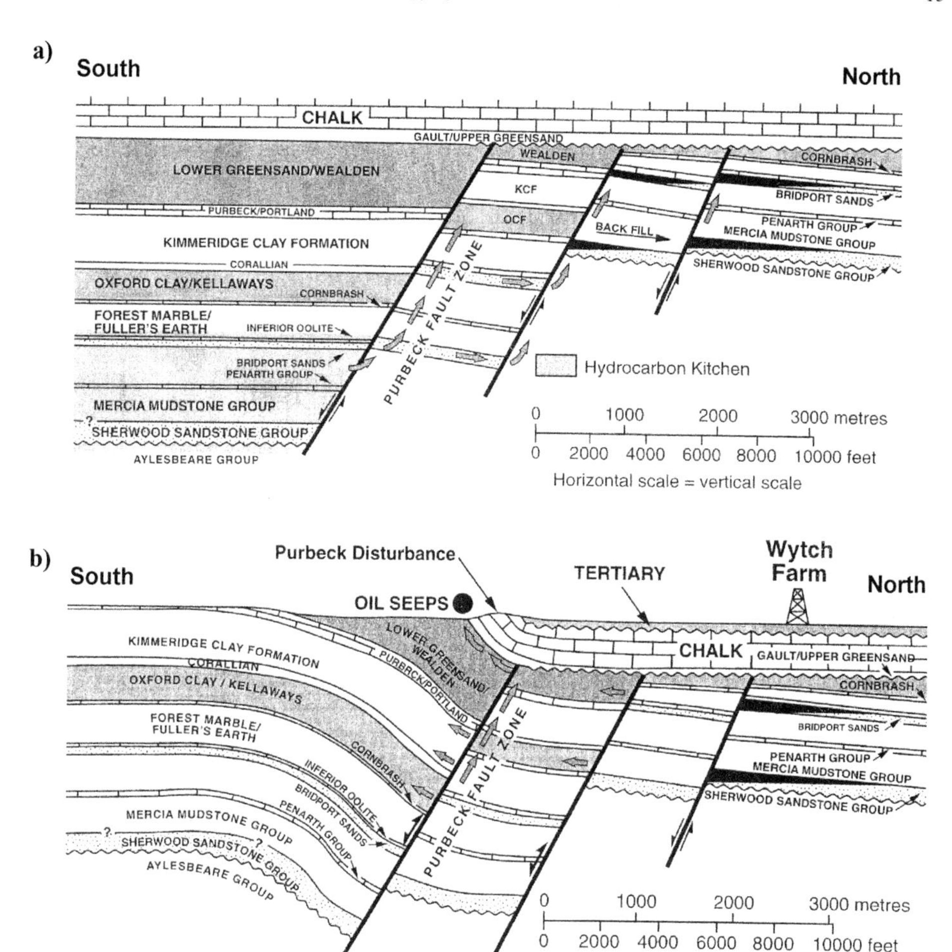

Figure 4 a) shows the reconstructed north-south cross-section of Wytch Farm at the time of oil migration in the Cretaceous. Extensional downfaulting to the south had taken the Lias source rock down into the hydrocarbon kitchen, the depth at which oil is generated. This migrated up fault planes into the Sherwood Sandstone and Bridport Sands beneath Wytch Farm. b) shows the present day cross-section with the position of Wytch Farm. Note that structural inversion has lead to the creation of the Purbeck Disturbance, and that oil which was probably present in a fault block to the south of Wytch Farm has been lost by leakage since the Cretaceous. From Underhill and Stoneley (1999) with permission.

warmest period in more recent earth history probably lead to melting of any polar ice which may have existed in the Antarctic. As a result, sea levels rose to the highest extent of Phanerozoic (post-Precambrian, the last 560 million years) time. The central part of continental Europe across to Central Asia was flooded, as was the Western Interior of the USA, and in these vast epicontinental seas Chalk was deposited. Chalk is a pelagic limestone formed from the calcite skeletons of myriad coccoliths, minute chrysophyte algae that live in the photic zone. Chalk was deposited more or less continually across the the UK for nearly 30 million years and it forms spectacular cliffs along the length of southern England from Devon to Kent. Chalk contains flint, derived by dissolution from the skeletons of siliceous organisms inhabiting the Chalk Sea.

Chalk has been important to the human species in various ways. The flint it contains proved ideal for the construction of tools and weapons by prehistoric peoples. The high porosity of many chalks has provided a fine aquifer, and although the Chalk has low matrix permeability, it has fractured extensively as a result of uplift, and the fracture zones form a conduit for water exploitation. The Chalk is the major aquifer across south Dorset and Hampshire.

The Alpine Orogeny: formation of the Hampshire Basin

The collision of Africa with southern Europe as the Tethys Ocean closed lead directly to the Alpine Orogeny and uplift of mountain chains which extend from Spain eastwards to Asia. The only existing remnant of the Tethys is the modern day Mediterranean. The compression from crustal shortening in the Alpine Orogeny was transmitted northwards and westwards across Europe, and resulted in reactivation of many deep structures. The Isle of Wight–Purbeck deep faults were no exception, and the infill of the Channel Basin was inverted, with development of steep dips on the strata immediately overlying the deep thrusts (Figure 4; Underhill and Paterson, 1998). Where the Chalk formed the nearly vertical northern limb of structures such as the Purbeck Anticline (and the Brixton and Sandown Periclines on the Isle of Wight), it partially recrystallized and created highly resistant ridges like the Purbeck Hills.

Although the Tethys Ocean started to close in the Late Cretaceous, as Africa collided with southern Europe, the first evidence of compression in the UK was the formation of a widespread unconformity at the summit of the Chalk, which probably dates to the earliest Palaeocene (Danian). Although this represents a considerable period of time (about 10 million years), the Chalk was only very gently folded. This unconformity was peneplaned by successive marine advances during the Palaeocene. Major inversion of structures in southern England commenced in the Mid Eocene, with initial uplift of the east-west periclinal structures in the Isle of Wight and Dorset, each of which overlie a deep basement thrust. The early history of this uplift is documented by reworking of progressively older sediments into the Hampshire Basin succession at Whitecliff Bay in the Isle of Wight during the Middle and Late Eocene (Gale *et al.*, 1999).

The major geomorphological consequence of this inversion was uplift of the Chalk ridges which run east-west through Dorset and the Isle of Wight, and formation of the Hampshire Basin to the north. The drainage pattern which is seen at the present day, with catchment from the west and north draining into the Solent along the axis of the Hampshire Basin, can thus be shown to have developed initially in the Palaeogene.

Eocene sedimentation and the proto-Solent

Poole Harbour is underlain by sands and clays of the Poole Formation; the Branksome Sand is restricted to the summit of Brownsea Island and the coast east of Poole (Figures 1 and 5). Both formations are of Middle Eocene age and were deposited in the broad coastal plain of the proto-Solent river complex which flowed eastwards, to the north of the Chalk hills of Purbeck and the Isle of Wight (Figure 6). It drained a catchment lying to the west and north. The position of the shoreline fluctuated considerably, but open marine conditions existed in the eastern Isle of Wight and West Sussex throughout much of the Eocene (Plint, 1988).

The Poole Formation comprises an alternation of clay units 5 m to about 30 m in thickness, and cross-bedded quartz sands which locally contain pebbles. The clays commonly display a fine alternation of sand and clay, on a millmetric to centrimetric scale, evidence of tidally dominated deposition on mudflats, supported by the presence of marine dinoflagellates. Five clay members (Creekmoor, Oakdale, Haymoor Bottom, Broadstone and Parkstone) have been named and mapped around Poole Harbour. The sands, which are commonly erosionally based and fine upwards were mostly deposited in fluvial channels, but some may be marine barrier sands. Current directions taken from cross bedding indicate a dominant source to the west, with minor northern and southern influences. The Branksome Sand is erosionally incised into the summit of the Poole Formation and comprises coarse to fine grained sands in fining upwards packages, deposited by rivers.

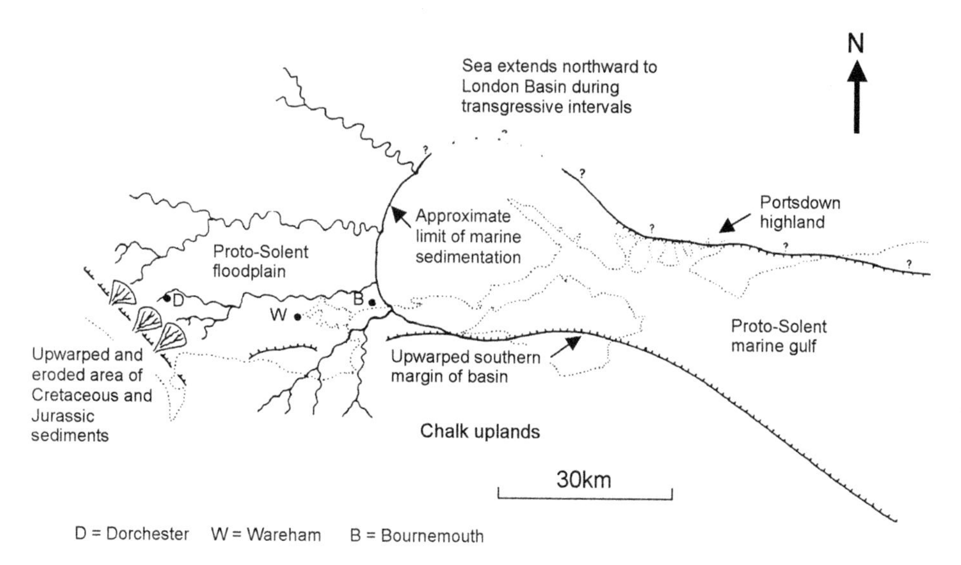

Figure 5 The succession in the London Clay, Poole and Branksome Formations in the Poole Harbour region. These sediments are of Mid Eocene age. The succession has been proved in many boreholes put down by English China Clay in the search for kaolinite deposits (ball clays). Modified after Bristow *et al.* (1991).

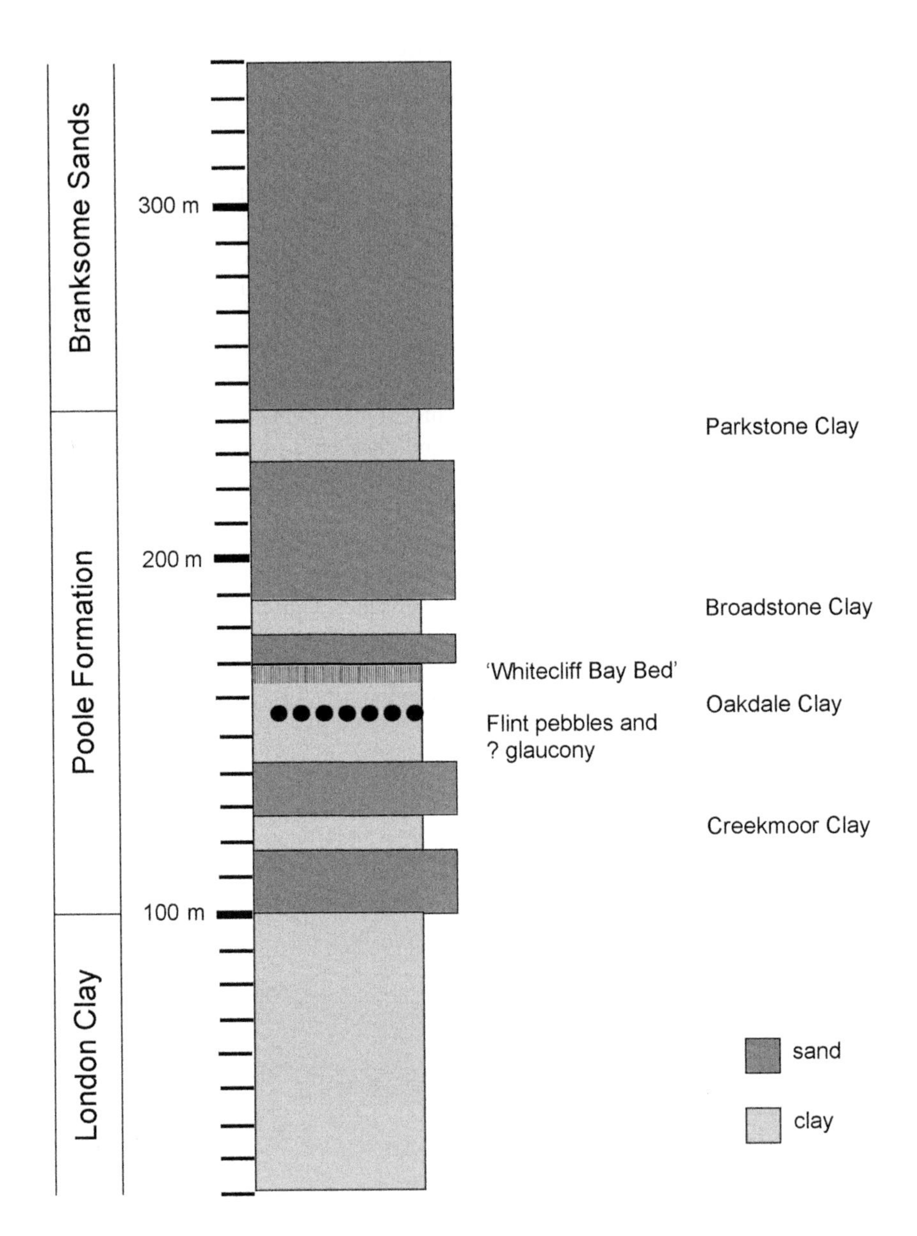

Figure 6 **Palaeogeographical map of east Dorset, south Hampshire and the Isle of Wight in the Mid Eocene. A gulf of the sea corresponding to the Proto-Solent extends from the east into the Hampshire Basin. Rivers from the west and north-west drain into this gulf, with more limited drainage from the Chalk uplands to the south and west. Poole Harbour was a region of fluviatile and brackish marine deposition.**

Additional evidence for the depositional conditions of the Poole and Branksome Formations comes from floras which are locally abundant in the laminated clays and are dominated by drifted leaves. Salt-tolerant ferns grew along the banks of rivers, adjacent to large expanses of sub-tropical forest. The clays themselves are of interest in terms of their ultimate sources, because the percentage of the clay mineral kaolinite increases westwards in the Poole Formation towards Wareham. Kaolinite is commonly derived from the weathering of granites in warm humid conditions, and the likelihood is that the catchment of the proto-Solent in the Eocene extended far westwards to the granite batholiths of Devon.

Evidence for correlation of the Poole and Branksome Sand Formations with the broadly coeval Bracklesham Group succession exposed in the Isle of Wight to the east is rather limited and inferential because neither formation yields fossils which provide high resolution ages. Although floras are locally diverse and abundant, only a limited number of rather long-ranging dinoflagellate taxa have been found, which indicates that most of the Poole Formation falls within the *coleothrypta* zone. The Parkstone Clay at the top of the formation and the overlying Branksome Sand fall within the *intricatum* zone above. Both zones represent long time intervals (probably several million years). Such correlations as have been made are essentially based on event stratigraphy, using in particular marine transgressions which have been identified in the Isle of Wight (Plint, 1988). A rooted palaeosol near the summit of the Oakdale Clay has been correlated with a rooted coal in the Wittering Formation of Whitecliff Bay in the east of the Isle of Wight, which may also be present in the Southampton district (Bristow *et al.*, 1991).

Because the transgressive events identified in the Isle of Wight are well marked by sharp facies changes, and were caused by eustatic rises in sea level of perhaps 10 m to 20 m, it seems eminently plausible that they have direct correlatives in the low-lying coastal palaeoenvironments represented by the Poole and Branksome Formations of east Dorset. The most marine succession in the island is that in Whitecliff Bay, which is broadly similar to that seen on the West Sussex coast on the Selsey peninsula (Figure 7). At this locality, four major transgressions can be identified in the Bracklesham Group, represented by fully marine shelly glauconitic sands which locally contain well-rounded black flint pebbles (T1–4 of Plint, 1988). These transgressive surfaces rest abruptly and erosionally upon tidally laminated sands and clays containing occasional rooted soils.

In the west of the island at Alum Bay, the succession is less marine and dominated by alluvial sands (Figure 7) (Plint, 1988). Nevertheless, the four transgressive events seen in Whitecliff Bay can be identified from the distribution of marine facies. The correlation can then be extended to the Poole district, taking the clay horizons in the Poole Formation as representing the most marine events. Thus, the trangression seen in the eastern Hampshire Basin within the Wittering Formation appears to correlate with the glauconitic green clay containing sparse flint pebbles in the Oakdale Clay of Poole Harbour (Bristow *et al.*, 1991).

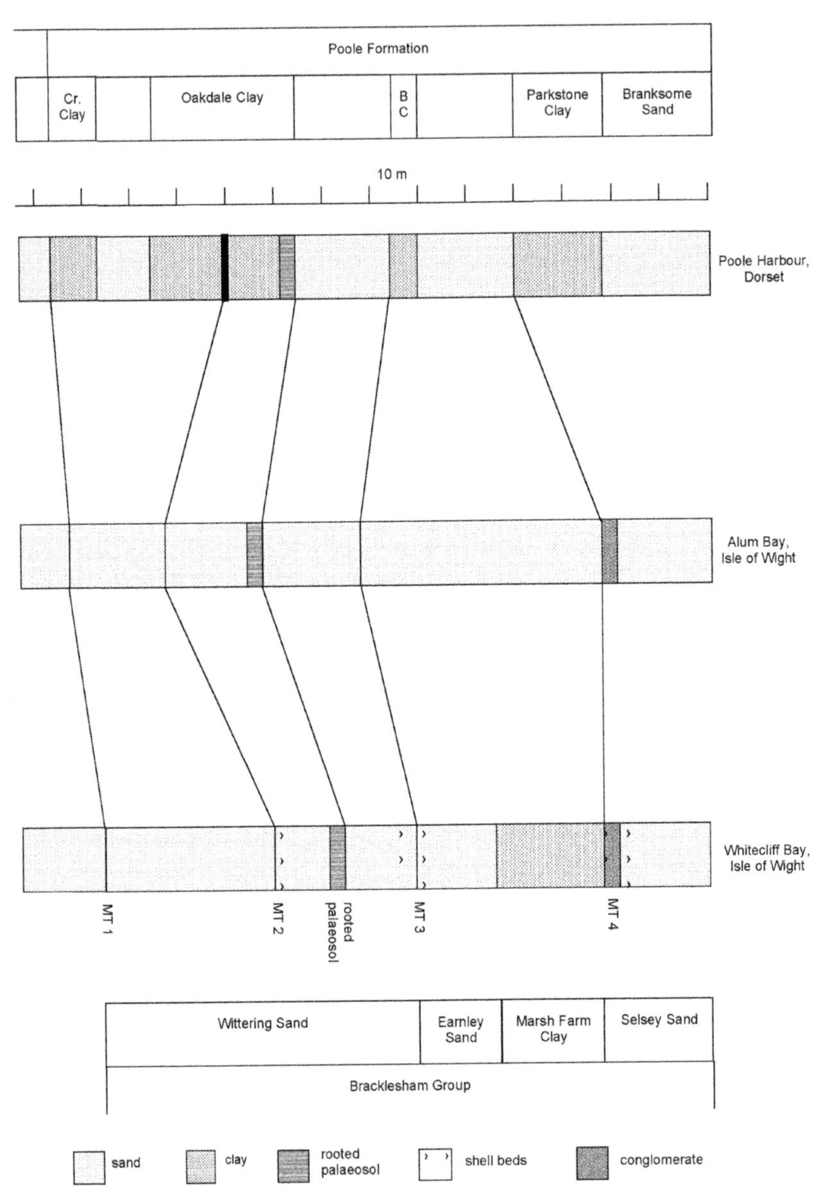

Figure 7 **Correlation of the Mid Eocene sediments between the Isle of Wight and the Poole Harbour region. This is based upon four marine transgressions (MT1–3) of which MT2–4 are represented in the eastern Isle of Wight (Whitecliff Bay) by marine shelly sands. In the Poole Formation, the transgressions are represented by the Creekmoor (Cr), Oakdale, Broadstone (BC) and Parkstone Clays, which were of restricted marine facies. A rooted palaeosol, called the Whitecliff Bay Bed, appears to correlate with a group of rooted palaeosols present in the summit of the Oakdale Clay. The localities are shown in Figure 1.**

The Palaeogene-Quaternary unconformity

In the Hampshire Basin, the Early Oligocene Solent Group is erosionally truncated (highest Bouldnor Formation, Isle of Wight) and there is no Late Oligocene or Miocene sedimentary record preserved here or elsewhere in southern England. Pliocene sediments are represented only by a few tiny outliers on the North Downs. The widespread unconformity between the Oligocene and Quaternary was interpreted by Wooldridge and Linton (1955) as evidence of a structural 'Alpine storm' of Miocene age, in which most of the anticlinal structures, such as those in the Weald and Purbeck–Isle of Wight, formed. However, this story is now replaced by one of more continuous uplift since the Mid Eocene when inversion seems to have commenced in the Isle of Wight at least (Gale *et al.*, 1999). This latter model is supported by other lines of evidence. Firstly, the high level of Pliocene deposits on the North Downs suggests that considerable uplift has taken place in southern England over the past few million years. Secondly, Preece *et al.* (1990) recorded uplift rates from the Late Quaternary of the eastern Isle of Wight which are almost identical to those calculated by Gale *et al.* (1999) for the Eocene succession. The ultimate cause of uplift, the collision of Africa and southern Europe, is likely to have continued fairly constantly over the last 50 million years, and still does so today.

Quaternary history

The drift deposits of Poole Harbour have been mapped (Bristow *et al.*, 1991) as river gravels, head and marine and estuarine alluvium. River gravels of Pleistocene age (the Frome Piddle Gravel Formation of Gibbard and Allen, 1995) directly overlie Middle Eocene sediments on the higher ground around Poole Harbour. These gravels were deposited by the easterly draining river complex of the Frome and Piddle, the major axial drainage system of the Hampshire Basin. These rivers originally drained into the substantial Solent River, which flowed eastwards along the line of the present Solent before eustatically driven marine erosion breached the Needles Channel during the early Holocene (Gibbard and Allen, 1995).

The river gravels comprise sandy gravels dominated by subangular flints and contain a small proportion of clasts including cherts and limestones of probable Jurassic age and sandstones derived from Tertiary sediments. The terraces have been numbered 1–14 in order of increasing age. The higher gravels are extensively cryoturbated from the effects of periglacial freeze-thaw. Gravels 1–10 have been identified in Poole Harbour (Bristow *et al.*, 1991). The deposits mapped as head comprise mostly gravels that have soluflucted as result of freeze-thaw action. The estuarine alluvium found extensively around Poole Harbour comprises a lower gravelly unit and an upper silty clay with a high organic content, each a maximum of 1 m in thickness. Both were deposited in the Holocene during the Flandrian transgression.

Acknowledgements

I would like to thank Angaharad Hills of the Geological Society Publishing house and John Underhill of Edinburgh University for permission to use figures from the paper by Underhill and Stoneley published in *Petroleum Geology of the Wessex Basin*, a Geological Society Special Publication. These are reproduced here in modified form as Figures 2 and 3.

References

Bristow, C. R., Freshney, E. C. and Penn, I. E. (1991) Geology of the country around Bournemouth. *Memoir of the Geological Survey of Great Britain*. Sheet No. 329.

Buchanan, J. G. (1998) The exploration history and controls on hydrocarbon prospectivity in the Wessex Basins, Southern England, UK. pp. 19–39. In: *Development, Evolution and Petroleum Geology of the Wessex Basin*. Underhill, J. R. and Stoneley, R. (eds). *Geological Society Special Publication,* No. 133.

Gale, A. S. (2000a) Late Cretaceous to Early Tertiary deposition through the global highstand. pp. 356–373. In: *Geological History of Britain and Ireland*. Woodcock, N. H. and Anderton, R. A. S. (eds). Oxford: Blackwell.

Gale, A. S. (2000b) The Cretaceous world. pp. 4–19. In: *Global Change and the Biosphere; The Last 150 Million Years*. Culver, S. J. and Rawson, P. F. (eds). Cambridge: Cambridge University Press.

Gale, A. S., Jeffery, P.A., Huggett, J. M. and Connolly, P. (1999) Eocene inversion history of the Sandown Pericline, Isle of Wight, southern England. *Journal of the Geological Society*, **156**: 327–339.

Hesselbo, S. P. (2000) Late Triassic and Jurassic: disintegrating Pangaea. pp. 314–338. In: *Geological History of Britain and Ireland*. Woodcock, N. H. and Anderton, R. A. S. (eds). Oxford: Blackwell.

House, M. R. (1993) *The Geology of the Dorset Coast. Geologists' Association Guide,* No. 22.

Gibbard, P. L. and Allen, L. G. (1995) Drainage and evolution in south and south-east England during the Pleistocene. *Terra Nova,* **6**: 444–452.

Plint, A. G. (1988) Global esutacy and the Eocene sequence in the Hampshire Basin. *Basin Research,* **1**: 11–22.

Preece, R. C., Scourse, J. D., Houghton, S. D., Knudsen, K. L. and Penney, D. N. (1990) The Pleistocene sea-level and neotectonic history of the eastern Solent, southern England. *Philosophical Transactions of the Royal Society of London, Series B*, **328**: 425–477.

Radley, J., Gale, A. S. and Barker, M. J. (1998) Derived Jurassic fossils from the Vectis Formation (Lower Cretaceous) of the Isle of Wight, southern England. *Proceedings of the Geologists' Association,* **109**: 81–91.

Ruffell, A. H. and Shelton, R. G. (2000) Permian to Late Triassic post-orogenic collapse, early Atlantic rifting, deserts, evaporating seas and mass extinctions. pp. 297–313. In: *Geological History of Britain and Ireland*. Woodcock, N. H. and Anderton, R. A. S. (eds). Oxford: Blackwell.

Selley, R. C. and Stoneley, R. (1987) Petroleum habitat in south Dorset. pp. 139–148. In: *Petroleum Geology of North-west Europe*. Brooks, J. and Glennie, K. (eds). London: Graham and Trotman.

Underhill, J. R. and Paterson, S. (1998) The structural development of the Wessex Basin. *Journal of the Geological Society of London.*

Underhill, J. R. and Stoneley, R. (1998) Introduction to the development, evolution and petroleum geology of the Wessex Basin. pp. 1–19. In: *Development, Evolution and Petroleum Geology of the Wessex Basin*. Underhill, J. R. and Stoneley, R. (eds). *Geological Society Special Publication,* No. 133.

Warr, L. N. (2000) The Variscan Orogeny: the welding of Pangaea. pp. 271–294. In: *Geological History of Britain and Ireland.* Woodcock, N. H. and Anderton, R. A. S. (eds). Oxford: Blackwell.

Wooldridge, S. W. and Linton, D. L. (1955) *Structure, Surface and Drainage of South-east England.* London: George Philip.

2. Geomorphology of Poole Harbour

Vincent May

School of Conservation Sciences, Bournemouth University, Talbot Campus, Fern Barrow, Poole, Dorset BH12 5BB

The geomorphology and sedimentation of Poole Harbour are poorly described although there are many localized studies within the harbour. Water level changes, both within the archaeological time-scale and at the present, may be a result not only of eustatic and regional isostatic changes but also of the effects of changes in tidal prism. Variations in marsh, intertidal areas and beaches have had important effects on the ecology of the harbour. Further investigation is needed of the landforms underlying both Sandbanks and the Studland dunes, as well as the responses of the intertidal areas and shore to changes in water levels.

Introduction

Poole Harbour is one of several estuaries in south-central England which are enclosed by spits and bars at their mouth and resulted from the drowning of valleys flowing into the English Channel. At High Water Spring Tides, its area is about 3600 ha, 1000 ha less than 6000 years ago. It has a single entrance and a long channel seawards between extensive sand shoals. There are three main channel networks. South Deep drains the southern lowland heaths, and the Wych Channel drains the Corfe River catchment. The northern harbour forms the estuary of the Rivers Frome and Piddle (with a combined catchment of over 770 km^2) and two smaller embayments with very restricted mouths, Lytchett Bay (the Sherford River estuary) and Holes Bay (draining the heathlands around Creekmoor). Much of the natural shoreline is marked by a low bluff (commonly less than 5 m in height) and eroding cliffs, but the northern shoreline is mainly artificial with walls, embankments, marinas and wharves. Tidal range is small (1.8 m at spring tides, 0.6 m at neaps at Poole Quay) and a double high water often holds water levels above mean tide level for 16 out of 24 hours. Tidal range increases upstream. Saltmarsh is confined to a relatively narrow fringe contained within a vertical elevation from around +0.05 m to +1.3 m Ordnance Datum (OD). The intertidal mudflats are not available for wading birds to feed on benthic invertebrates for extensive periods. The harbour has relatively poor flushing characteristics.

The geomorphology of Poole Harbour is poorly described. Most investigations focus on the accumulation and release of sediment associated with the spread and dieback of *Spartina anglica* (Bird and Ranwell, 1964; Gray, 1985; Gray *et al.*, 1995; Raybould, 1997), sedimentation and dredging in the main navigable channels (Green, 1940; Green *et al.*, 1952; Halcrow Maritime, 1999) and the development of the harbour mouth beaches and dunes (Diver, 1933; Robinson, 1955; Carr, 1971; May, 2003). Investigations of the Holocene sediments typically focus on the vegetational history (Haskins, 1978) or interpretations of sea level change (Nicholls, 1986; Edwards, 2001, Long *et al.*, 1999).

The morphology of the harbour and its entrance

The morphology of the harbour is dominated by intertidal banks and slopes and sub-tidal channels mostly in loose sediments resting on the rarely exposed underlying Poole Formation (Fahy *et al.*, 1993). The channel depth at the harbour mouth is about -15 m LAT (about -16.4 m OD). However, the depth of bedrock in the Swash Channel off Shell Bay is about -10.6 m OD (BP Exploration, 1991). East of the Hook Sands, a channel about 1 km wide attains depths in bedrock of about -14 m OD (BP Exploration, 1991).

The landward edge of much of the harbour is marked by:

- a low bluff typically less than 5 m in height or
- actively or partly degraded cliffs up to 10 m in height or
- artificial structures (May, 1969).

To seaward, there are extensive saltmarshes and mudflats or small sand and shingle beaches. The highest level of active wave erosion of cliffs varies in height (between about -0.6 to -1.1 m OD). However, at Shipstal, most cliff foot retreat occurs during periods of high wave-runup to above High Water Springs. Shore platforms cut into bedrock typically have a narrow upper slope of about 1 in 8 and then a wider slope seawards at angles of between 1:80 and 1:120. Beneath the saltmarsh, there is occasionally a step about 0.5–0.6 m high between -0.6 m OD and -1.1 m OD (Edwards, 2001). The saltmarsh surface typically falls in the southern harbour from about +1.3 m OD to +0.05 m OD, where it is truncated by an eroding cliff about 0.2– 0.3 m in height.

Tides, waves and sea level rise in Poole Harbour

The geomorphology of the harbour results from the combined effects of:

- marine and sub-aerial processes on both intertidal zone and shoreline
- the channel hydrodynamics
- anthropogenic modifications of the shoreline and channels
- catchment hydrology affecting both the freshwater and sediment inputs
- the spread and decline of the saltmarshes.

Sea level rise, tides and waves affect the estuary on different time-scales, the effects of waves varying with the tidal cycle and water levels (Table 1). Low-angle intertidal slopes within the harbour mean that these combined effects may be detectable over the time-scale of the *Spartina* era.

Table 1 Estimates of sea level altitude and change

(a) Changes up to late nineteenth century

Location	Dates	Altitude (m OD)	Source	Period and direction of change	Years	Mean annual change (mm $year^{-1}$) derived from sources
Hamworthy	7340 +/ 110 BP	-12.8	Godwin *et al.* (1958)	Rising 7340–6000 BP	1340	+9.6
Poole Harbour	6000 BP	Present level	Bird and Ranwell (1964)	Stable with slight falls		
Arne and Newton Bay			Edwards (2001)	Rising 4700–2400 BP	2300	+0.26–0.7
Arne and Newton Bay			Edwards (2001)	Falling/stable 2400–1200 BP	1200	
Brownsea	End third century AD	- 2.7	Jarvis (1992)	Rising 1600 to present	1600	+1.7
Arne and Newton Bay			Edwards (2001)	Rising then stable 1200–900 BP	300	
Arne and Newton Bay			Edwards (2001)	Rising 400–200 BP	200	
Arne and Newton Bay			Edwards (2001)	Net rise 4700–200	4500	+0.53

(b) Post nineteenth century changes

Location	Dates	Altitude (m OD)	Source	Period and direction of change	Years	Mean annual change (mm $year^{-1}$) derived from sources
Keysworth	1912–1966		Hubbard and Stebbing (1968)	Rising	54	+1.85
Poole	Current		Pethick (1993)	Rising		+4.0

Although close to present levels for the last 6000 years, sea level has been lower during that period (Table 1a), rising and falling within a vertical range of about 3 m (Edwards, 2001; Long *et al.*, 1999). During this period, the rate of change in sea level was significantly lower than in either the preceding period of very rapid rise or in the present much shorter period of accelerated sea level rise. The altitude of the marshes and shoreline erosion also depend on variations in wave heights and tidal range. Hubbard and Stebbing's (1968) estimate suggests that sea level would have risen by 170 mm between 1912 and 2004. However, recent estimates imply that sea level rise since the mid-1960s has been faster. Combining these two estimates indicates a net sea level rise since the arrival of *Spartina* of about 0.28 m. Hubbard and Stebbing (1968) give High Water Spring Tides at +0.79 m OD, but Halcrow Maritime (1999) has Mean High Water Springs at +0.9 m OD, a difference equivalent to an annual rise of 3.5 mm between 1966 and 1997, i.e. consistent with Pethick's (1993) value. Given the low angle slopes of the intertidal areas (often flatter than 1:100), such a rise of water levels has implications for marsh edge beaches and cliff retreat, as well as the plant communities around the estuary.

The tidal range at Poole North Haven is 1.7 m (Halcrow Maritime, 1999). Green (1940) showed that the tidal peak not only lagged upstream, but also the tidal curve steepened upstream during the flood tide. Ranwell *et al.* (1964) estimated the tidal range at Keysworth as 2.9 m. Edwards (2001) in contrast has tidal ranges of 1.2 m and 1.6 m at Arne and Newton Bay, respectively. The tidal prism, i.e. the volume of water exchanged on each tidal cycle, is affected by the shape of the estuary. If land claim or accretion occur, the intertidal volume is reduced and estuaries adjust by changing the shape of the estuary mouth. Poole, however, has an artificially narrow mouth and so reduced tidal prism is likely to cause water levels to rise and the sub-tidal channels to deepen (Table 2). Varying water levels during the past century may result from these local conditions rather than sea level rise alone. Tidal currents attain 2.0 m s^{-1} in the entrance channel, but are generally slower in the main channel: 0.5 m s^{-1} (Halcrow Maritime, 1999). Ebb-dominated in terms of the currents, the estuary as a whole is wave-dominated in terms of the overall energy distributions, but there is a contrast between the ebb-domination of the lower harbour and the flood-dominated upper estuary (Green, 1940).

Wind direction and strength and fetch affect wave formation. Refracted English Channel waves affect the harbour mouth and open-coast beaches, but do not penetrate the majority of the harbour. Winds of about 15 knots will generate waves with significant wave height H_s 0.38 m over a fetch of 5 km (the distance from Parkstone Bay to Shipstal Point) and similar winds blowing from WNW would generate waves with H_s 0.40 m at Shore Road.

Geomorphological history of the harbour

The harbour is cut into the Bracklesham Group, mostly the Poole Formation, an "alternating sequence of fine- to very coarse-grained, locally pebbly, cross-bedded sands, and pale grey to dark brown, carbonaceous and lignitic, commonly laminated

Table 2 Key hydrodynamic and morphological changes during the *Spartina* era

	Saltmarsh area	Tidal prism	Sub-tidal channel depth	Tides and currents	South Deep
c.1880–c.1925	Increase area = 775 ha by 1925	Decrease c. 20%	Increase	Intertidal elevation increase	50% deeper
1925–present	Dieback area = 415 ha in 1980	Increase c. 11%	Decrease	Increased ebb currents at harbour mouth	20% shallower

Source: Based on Gray *et al.* (1995).

clays" (Bristow *et al.*, 1991, p. 33). The earliest preserved geomorphological landscape comprises valleys and terraces associated with the Solent River (Everard, 1954). River Terrace Deposits, typified by gravel and sand deposits up to 3 m thick, occur at altitudes up to +50 m OD.

As sea levels fell, the rivers cut channels to at least -13 m OD, forming a landscape of rivers meandering between low hills and ridges separating the northern estuary from the south. Wright (1982) suggested that a southern re-orientation of the drainage in Poole Bay began in the early Devensian. Velegrakis (1994) argued that an 'English Channel' river flowed directly southwards. Tyhurst and Hinton (2004) suggested that Poole Bay was deeper than Christchurch Bay for much of the late Holocene, arguing that, as the bays flooded and the coastal configuration changed, tidal range would also alter. The nature and rate of coastal changes would thus differ from today.

As Holocene sea level rose, the valleys flooded, establishing the present pattern of deep channels. Sea level probably attained present levels about 6000 years ago, but archaeological investigations suggest lower sea levels both during the Iron Age (Wilkes, pers. comm.) and in the late third century AD (Jarvis, 1992). With mean tide level (MTL) about -2.7 m OD, but assuming that tidal range has not changed, erosional retreat processes would occur along the cliffed coast at about -1.8 m OD. This appears consistent with the step under Arne (Edwards, 2001). MTL does not always represent the shoreline or the level at which erosion processes will be concentrated, i.e. the underlying bedrock surface may slope upwards as it does today. Moreover, assuming that wind strengths and storm events were similar to today, most direct erosion of the shore would occur below MTL. The presence of sand layers in corings at Arne and Newton Bay (Long *et al.*, 1999; Edwards 2001) does not necessarily imply an erosional phase, since within the harbour such deposition of sandy deposits also depends upon transport patterns, the supply from existing cliffs and the presence of marsh surfaces on which deposition can occur. Phases of greater wave energy or shifts in wind direction (and so wave climate) may also produce areas of deposition on existing marshes.

For the past six millenia, channels, islands and intertidal ridges and flats have characterized the harbour. Although the harbour mouth is well established, the point at which the harbour flooded during the period of rapid sea level rise is not certain. The Godwin *et al.* (1958) dating for marine sediments overlain by freshwater peat at Hamworthy would put sea level about 7500 years BP about -12.5 m OD. However, with the bedrock altitude in the Swash Channel at -10.6 m OD, the sea could not enter the harbour by this route. The deeper bedrock channel east of the Hook Sands (and coincidentally opposite the narrow neck of Sandbanks) lies at about -14 m OD. If this channel extends landwards (for which currently there is no direct evidence), this would provide evidence of a separate entrance to the northern harbour between Sandbanks and Flag Head. This needs investigation.

Finally the arrival of *Spartina anglica*, large-scale reclamation around Poole and the regular maintenance and capital dredging of the main channels have altered the detailed shape of the estuary. Much of the northern shore has been reclaimed, the overall area of the estuary has reduced and the tidal prism has been restricted.

Recent sedimentation and change within the harbour

There was little change in patterns of channels, mudflats and fringing marshes until the end of the nineteenth century (Table 3). *Spartina* growth captured about 7 million m^3 of sediment by 1925, but as saltmarshes died back, at least 4 million m^3 was released (Gray *et al.*, 1995).

Most marsh sediments are fluvial in origin. The embayments are typically areas of accretion, often reclaimed and so trapping suspended sediments. Cores and boreholes

Table 3 Recent trends in harbour hydrodynamics

- Under 3 million m^3 fine sediment still retained in saltmarsh but reducing
- Channel width at harbour mouth fixed and channel scoured and dredged
- Tidal prism not yet returned to pre-*Spartina* level
- Tidal prism reduced by land claim
- Sub-tidal channels deeper than pre-*Spartina* but shallowing in upper estuary
- Approximately 0.26 m water level rise since late nineteenth century

vary from 3.0 m to 21.3 m deep before the underlying Poole Formation strata are reached (Bristow *et al.*, 1991). Most contain layers of peat. The deepest at Ferry Road (SZ 008 902) identified bedrock at 21.3 m *below the surface.* 'Oyster' shells at depths of up to 7.3 m *below the surface* were underlain by shelly soft clay changing downwards to silty sand. At SZ 00679007, a 1.0 m layer of silt containing peat occurred at 8.3 m and the Poole Formation at 11.3 m *below the surface.*

Sedimentation was rapid under the influence of *Spartina*. At Keysworth seaward of the reedswamp, Hubbard and Stebbing (1968) reported marsh sediments 1.98 m thick. The muddy upper marsh lay between +0.77 m OD and -0.84 m OD, with over 0.95 m sand above 0.1 m peat resting on gravels and Bagshot Beds at –1.21 m OD. In contrast, the lower central marsh sediments were over 6.25 m thick. Following *Spartina* colonization in 1912, the mean annual accretion rate was 32 mm. However, annual rates were greater (e.g. 80 mm in 1963; up to 150 mm in 1966). Total post-*Spartina* accretion was estimated at over 180 cm at Keysworth, 70 cm in Arne Bay and 35 cm in Brands Bay (Bird and Ranwell, 1964; Ranwell, 1964; Hubbard and Stebbing, 1968). There are few measures of sediment input from the rivers. Hubbard and Stebbing (1968) give a sample estimate of suspended matter at Wareham Quay of 36,700 kg per day but much of this may be organic. May (1969) noted that the annual provision of sediment from the rivers was limited. Changed agricultural practices in the catchment may have altered the patterns of soil erosion, particularly in the autumn.

Maximum salinity coincides with minimum sediment concentration at the junction of the Main and Middle Channels (Green, 1940). The channels, except for Main Channel, only shifted position slightly during the nineteenth and early twentieth centuries. Main Channel had phases of movement towards the east and north-east between which it moved little but became shallower. The sediments of the estuary are generally fine (Green *et al.*, 1952), but relatively coarse in the area between Brownsea and Salterns. Suspended sediment distributions have been modelled by Falconer (1984) and Falconer and Chen (1991) using finite differences solutions to predict time varying water elevations, velocity, bed shear stresses and suspended and bed load sediment flux concentrations. For Poole harbour as a whole (Pethick and Forster, 1994), the highest values occur in the entrance channel, in the main channel east of Brownsea and in the channels between the islands.

The beaches and cliffs within the harbour

Cliff erosion provides the many small beaches with gravel and sand. Gravels often form layers on the upper part of the intertidal flats. Some gravel is rafted onshore attached to wracks. Beaches inside the harbour typically show a cliff: cliff-foot beach: spit pattern, often with a post-*Spartina* chenier on the marsh edge. There are also some larger sandy spreads especially between the islands. Before *Spartina* colonized, the islands and headlands, such as Baiter, commonly had a spit extending eastwards along the southern shore and recurving northwards at the eastern end of the island. On the northern side of the islands and peninsulas, fringing beaches with small ness-like areas of sand and gravel beach ridges (e.g. Seymours on Brownsea Island) are more common. After *Spartina* spread, marsh-edge beaches become more common.

Cliff erosion affects the northern shore and the islands. The estimated annual retreat rate of the cliffs during the nineteenth and early twentieth centuries was 0.4 m (May, 1969). The cliffs on Green Island have probably retreated about 170 m since the Iron Age (i.e. 0.06 m year^{-1}) . For much of the twentieth century, *Spartina* protected existing cliffs, retreat rates fell to about 0.1 m year^{-1} and many former cliffs degraded to become well-vegetated slopes: e.g. at Portland Hill, Brownsea Island, retreat of the shoreline averaged 0.1 m year^{-1} pre-*Spartina*, fell to 0.02 m year^{-1} with *Spartina* protection and then increased to 0.85 m year^{-1} when saltmarsh eroded. The shelter afforded by the islands has decreased with sea level rise and cliff erosion. Parts of the cliffs at Goathorn are retreating at present and earlier features are being reworked.

The cliffs and beaches at Shipstal, where a spit is first recorded in 1785, demonstrate the complexity of interaction between changes in wave energy distributions, cliff retreat and sediment supply and transport, the role of saltmarsh and changes in water levels. Nineteenth century beach development was limited (May, 1976), but the spread of *Spartina* in Arne and Middlebere bays altered local wave patterns and provided a base for extended beaches. A new beach formed southwards and the northern beach extended into Arne Bay into deeper water, and on to looser sediments. May (1981) predicted, based on an investigation of the sediment budget, that the historical pattern would change because of three linked factors: a reduced supply of sediment from the south, increased frequencies of north-easterly winds, and a decline of the proximal area of the beach and an increase in overwash. Since the 1980s, this pattern has dominated, but erosion of the proximal beach has increased, so that the older (eighteenth and nineteenth century) ridges eroded by about 20 m since 1970.

The beaches and mouth of the harbour

The beaches and dunes at the mouth of Poole Harbour have been the focus of investigations for many years. In particular, the development of the dunes on the South Haven peninsula has been well documented (see May, 2003, for summary) and so will not be discussed further. Investigations by Halcrow Maritime (1999, 2004) analysed contemporary processes in detail, largely corroborating earlier studies.

The main source of sand appears to be the seabed: extensive sand banks occur throughout the cartographic record. It is unclear why there was a sudden onset of dune building in the seventeenth century (Diver, 1933), and there is no explanation of the wide intertidal area forming their base. The early history of the dunes is, therefore, far from clear, although it is certain that the dunes at Studland have an earlier origin than Diver or later writers suggested (May, 2003). Both the sub-dune surface and the sedimentation processes on South Haven need further investigation.

References

Bird, E. C. F. and Ranwell, D. S. (1964) *Spartina* salt marshes in southern England. IV. The physiography of Poole Harbour, Dorset. *Journal of Ecology*, **52**: 355–366.

BP Exploration (1991) *Hook Island – Poole Bay: Private Bill – Environmental Study.*

Bristow, C. R., Freshney, E. C. and Penn, I. E. (1991) *Geology of the Country Around Bournemouth. Memoir 1:50,000 Geology Sheet 329 (England and Wales).* London: HMSO.

Carr, A. P. (1971) South Haven Peninsula: physiographic changes in the twentieth century. pp. 32–38. In: *Captain Cyril Diver, a Memoir*. Merrett, P. (ed.). Nature Conservancy Council Furzebrook Research Station, Nature Conservancy Council.

Diver, C. (1933) The physiography of South Haven Peninsula, Studland Heath, Dorset. *Geographical Journal,* **81**: 404–427.

Edwards, R. J. (2001) Mid- to late Holocene relative sea-level change in Poole Harbour, southern England. *Journal of Quaternary Science*, **16**: 221–235.

Everard, C. E. (1954) The Solent River: a geomorphological study. *Transactions of the Institute of British Geographers*, **20**: 41–58.

Fahy, F. M., Hansom, J. D. and Comber, D. P. M. (1993) *Estuarine Management Plans – Coastal Processes and Conservation in Poole Harbour.* Glasgow: Coastal Research Group, Department of Geographical and Topographic Sciences, University of Glasgow.

Falconer, R. A. (1984) *A Mathematical Model Study of the Flushing Characteristics of a Shallow Tidal Bay*. Report for English Nature.

Falconer, R. A. and Chen, Y. (1991) An improved representation of flooding and drying and wind stress effects in a two-dimensional numerical model. *Proceedings of the Institute of Civil Engineers, Part 2, Research and Theory*, **91**: 659–678.

Godwin, H., Suggate, R. P. and Willis, E. H. (1958) Radiocarbon dating of the eustatic rise in ocean-level. *Nature*, **181**: 1518–1519.

Gray, A. J. (1985) *Poole Harbour: Ecological Sensitivity Analysis of the Shoreline*. Abbots Ripton: Institute of Terrestrial Ecology.

Gray, A. J., Warman, E. A., Clarke, R. J. and Johnson, P. J. (1995) The niche of *Spartina anglica* on a changing coastline. *Coastal Zone Topics: Process, Ecology and Management*, **1**: 29–34.

Green, F. H. W. (1940) *Poole Harbour: A Hydrographic Survey 1938–1939*. London: Geographical Publications Ltd for Poole Harbour Commissioners and University College, Southampton.

Green, F. H. W., Ovington, J. D. and Madgwick, H. A. I. (1952) Survey of Poole Harbour: changes in channels and banks during recent years. *The Dock and Harbour Authority*, **33**: 142–144.

Halcrow Maritime (1999) *Poole and Christchurch Bays Shoreline Management Plan.* Swindon: Halcrow Maritime.

Halcrow Maritime (2004) *Poole Bay and Harbour Strategy Study – Assessment of Flood and Coast Defence Options*. Final Report for Bournemouth and Poole Borough Councils –February 2004. Swindon: Halcrow Maritime.

Haskins, L. E. (1978) The Vegetational History of South-east Dorset. Unpublished PhD thesis, University of Southampton.

Hubbard, J. C. E. and Stebbing, R. E. (1968) *Spartina* marshes in southern England. VII. Stratigraphy of Keysworth marsh, Poole Harbour. *Journal of Ecol*ogy, **56**: 702–722.

Jarvis, K. (1992) An intertidal zone Romano-British site on Brownsea Island. *Proceedings of the Dorset Natural History and Archaeological Society,* **117**: 89–95.

Long, A. J., Scaife, R. G. and Edwards, R. J. (1999) Pine pollen in intertidal sediments from Poole Harbour, UK; implications for late-Holocene sediment accretion rates and sea-level rise. *Quaternary International*, **55**: 3–16.

May, V. J. (1969) Reclamation and shoreline change in Poole Harbour, Dorset. *Proceedings of the Dorset Natural History and Archaeological Society,* **90**: 141–154.

May, V. J. (1976) Cliff erosion and beach development, Shipstal Point. *Proceedings of the Dorset Natural History and Archaeological Society*, **97**: 8–12.

May, V. J. (1981) *Assessment of the Impact of Changes in the Environment of Poole Harbour Upon Its Upper Reaches*. Poster presentation for Heritage Education Project Conference and Exhibition 'Now and Then', 3 November 1981, Wareham.

May, V. J. (2003) South Haven Peninsula. pp. 340–345. In: *Coastal Geomorphology of Great Britain*. May, V. J. and Hansom, J. D. (eds). *Geological Conservation Review Series*, No. 28. Peterborough: Joint Nature Conservation Committee.

Nicholls, R. J. (1986) The evolution of the upper reaches of the Solent River and the formation of Poole and Christchurch bays. pp. 99–114. In: *Wessex and the Isle of Wight: Quaternary Research Association Field Guide*. Barber, K. E. (ed.).

Pethick, J. (1993) *Cliff Erosion: Furzey Island, Poole Harbour*. Report to BP Environmental. Hull: Institute of Estuarine and Coastal Studies, University of Hull.

Pethick, J. and Forster, A. (1994) *Furzey Island: Cliff Erosion*. Report to BP Environmental. Hull: Institute of Estuarine and Coastal Studies, University of Hull.

Ranwell, D. S. (1964) Spartina saltmarshes in Southern England: II. Rate and seasonal pattern of sediment accretion. *Journal of Ecol*ogy, **52**: 79–94.

Ranwell, D. S., Bird, E. C. F., Hubbard, J. C. E. and Stebbings, R. E. (1964) Spartina saltmarshes in Southern England: V. Tidal submergence and chlorinity in Poole Harbour. *Journal of Ecol*ogy, **52**: 627.

Raybould, A. F. (1997) The history and ecology of *Spartina anglica* in Poole Harbour. *Proceedings of the Dorset Natural History and Archaeological Society,* **119**: 147–158.

Robinson, A. H. W. (1955) The harbour entrances of Poole, Christchurch and Pagham. *Geographical Journal*, **121**: 33–50.

Tyhurst, M. F. and Hinton, M. T. (2004) *The Evolution of Poole and Christchurch Bays: Another Look at the Flandrian Transgression*. (Accessed at www.christchurch.gov.uk/I%26t/coastpro/evoluc.pdf)

Velegrakis, A. F. (1994) Aspects of the Morphology and Sedimentology of a Transgressional Embayment System: Poole and Christchurch Bay, Southern England. Unpublished PhD thesis, University of Southampton.

Wright, P. (1982) Aspects of Coastal Dynamics of Poole and Christchurch Bays. Unpublished PhD thesis, University of Southampton.

The Ecology of Poole Harbour
John Humphreys and Vincent May (editors)

3. Salinity and Tides in Poole Harbour: Estuary or Lagoon?

John Humphreys

University of Greenwich, Old Royal Naval College, Greenwich, London SE10 9LS

Notable features of Poole Harbour include its double high water and small tidal range. Due to the mixing of sea water with fresh water, surface salinity maxima and minima decrease while salinity range increases from the harbour entrance to its upper reaches. As far up as the Wareham Channel, salinity varies on a semi-diurnal basis in line with the tidal cycle. Salinity variation with depth is most pronounced in the Wareham Channel which is categorized as partially mixed, while the main body of the harbour is more or less vertically homogeneous. Depending on location, interstitial water in the mudflats may be of higher or lower salinity than the overlying water at high tide. Assertions in the literature that Poole Harbour is a lagoon are examined in the context of its tidal and salinity regimes. It is concluded that it is sufficient in explanatory terms to regard the harbour as an estuary.

The tidal range

The coastal waters of north-west Europe rise and fall twice a day under the dominant gravitational influence of the moon. Superimposed on this lunar tidal cycle is a lesser solar tide and while both are semi-diurnal, they are not of exactly the same period, the lunar day being 50 minutes longer than the solar day. Every 2 weeks, however, around the time of the new and full moon, the two tidal cycles are synchronized and spring tides occur which are of relatively large amplitude. Halfway between spring tides, the solar and lunar effects are opposed to produce reduced amplitude neap tides. So while the times of high and low water correspond with the lunar period, the predicted level to which tides rise and fall are determined over a 28 day period by an interaction of lunar and solar effects. In practice, however, whilst these gravitational cycles are relatively simple, actual tidal behaviour is more complex, due to other astronomical cycles associated with the seasons and weather conditions.

Were Poole Harbour to be an isolated water mass, its size is such that direct gravitational effects on it would be negligible. In fact, the tides of the harbour are transmitted into it from the adjacent sea. The coastal waters of Britain are set in oscillation by the rise and fall of Atlantic tides, and the English Channel effectively represents a basin in which the tides can be likened to a standing wave exhibiting greater amplitude in some locations with very little in others. The latter component of the standing wave is referred to as the nodal line. In the sea, the nodal line is reduced to a point or 'amphidrome' due to the rotation of the earth. In the northern hemisphere, the tidal oscillation rotates anti-clockwise round the amphidrome and tidal range increases with distance from it.

In contrast to the North Sea, there are no amphidromes in the English Channel. However, the flood tide does show a tendency to rotate anti-clockwise as it progresses west to east from the Atlantic. This behaviour has been described in terms of a degenerate amphrodromic system, the tide acting as if there would be an amphidrome inland of Bournemouth if the sea extended that far (Green, 1940; Tait, 1981). Consistent with this is the conspicuous contrast between the spring tide range at Poole Bay (around 2 m) and on the opposite French coast at St Malo (around 12 m). As a consequence, the tides in Poole Harbour are also of small amplitude, around 1.8 m for spring tides and only 0.6 m for neaps. Poole Harbour is, therefore, categorized as microtidal, i.e. of tidal range less that 2 m (McClusky and Elliot, 2004). Despite this, it is estimated that during a spring tide ebb up to 71.346 x 10^6 m^3 of water leaves the harbour, this being around 45% of the total spring high water volume (i.e. tidal prism ratio for spring tides = 0.45). For neap tides, 22% of the neap high water volume leaves on the ebb (Riley, 1995; Falconer, 1984). The geomorphology is such that around 80% of the harbour bed falls within the intertidal zone (i.e. between the extreme high and extreme low water levels for spring tides) and is, therefore, exposed periodically by the ebb tide (Gray, 1985).

The double high water

In a low amplitude tidal regime, local effects which would otherwise be masked become relatively pronounced. Consequently, on that part of the south coast of England between Portsmouth and Lulworth, a particular local tidal phenomenon known as a double high water occurs. This feature is transmitted into Poole Harbour. Figure 1 shows typical spring and neap tide cycles at Poole Bridge. The main low water is followed by a main flood tide which generates the first high water. A 'fore-ebb' generates a subsidiary low water, which is brought back up by a second flood tide (the 'half-flood') to a second high water which precedes the next main ebb and main low water. It can also be seen from Figure 1 that for spring tides the first high water is higher than the second, while for neap tides the reverse is true. It appears that the spring pattern is more common. Green (1940) reported that of 27 tides measured between April and May, in only seven was the second high water higher than the first. In any event, the detail of tidal behaviour in the harbour is more unpredictable than such generalizations imply, not least due to the effects of wind and barometric pressure. The double high is, however, a consistent and significant phenomenon.

Because of the double high tide, there is in Poole Harbour always a relatively long stand of high water such that for about 16 out of 24 hours, the water level is above mean tide level. This is of ecological significance as it limits the availability of mudflats as feeding grounds for important wader populations, while conversely increasing the feeding time for many of the filter feeding invertebrate animals of the mudflats which contribute to the diet of waders and also provide local fisheries. Effects on the distribution of saltmarsh plants have also been reported (Ranwell *et al.*, 1964).

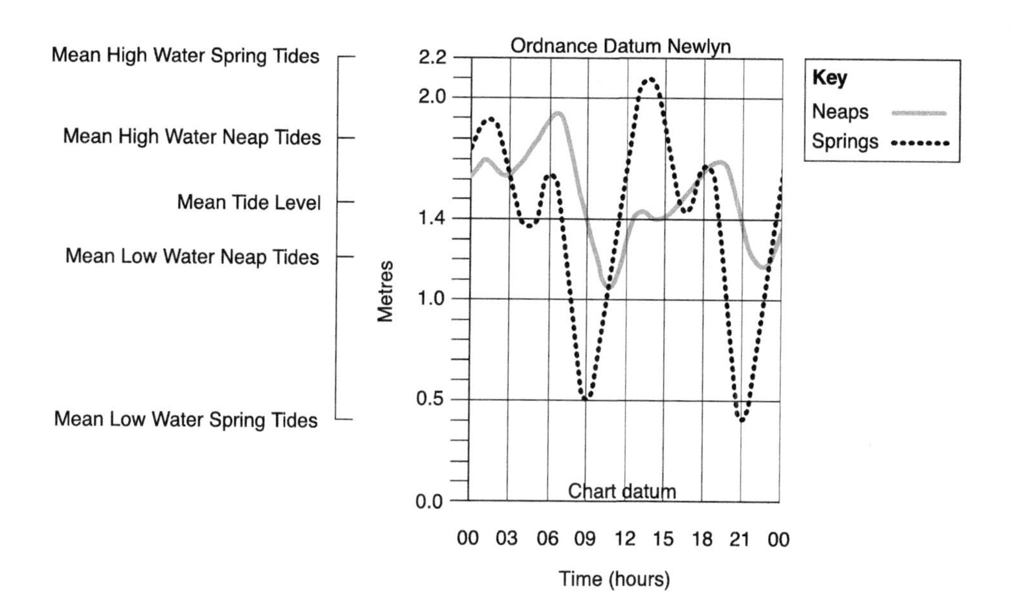

Figure 1 Typical tidal curves at Poole Bridge (modified from Gray, 1985).

Salinity

Classic definitions of salinity refer to the amount of dissolved inorganic matter expressed as grams per kilogram of water. In the open ocean, the salinity of sea water approximates to 35 g kg^{-1}. Normally, over 30 of the 35 g kg^{-1} of inorganic matter is made up of sodium and chloride – these being the main chemical components of salt. In contrast, the salinity of river water is generally less than 0.5 g kg^{-1} and the mixing of fresh water with sea water as a river approaches the sea is a defining feature of estuaries (Pritchard, 1960).

For aquatic organisms the salinity of the water in which they live is of considerable physiological significance. Most marine taxa have body fluid concentrations essentially the same as the surrounding sea water, consequently there is no tendency for them to gain or lose water by osmosis. Such osmotic equilibrium is not a feature of freshwater organisms, which in order to survive must maintain their internal concentration above that of the surrounding water. In this sense, fresh water is a more physiologically demanding medium than sea water, with the consequence that a lesser number of species have yet achieved the adaptations necessary to live in it.

Notwithstanding these challenges, rivers at least have the merit of being relatively stable in terms of the concentration of the water – a feature that freshwater environments share with the open sea. In contrast, the mixing of river and sea water in an estuary creates an environment of peculiar severity in which salinity varies considerably on seasonal and indeed shorter time-scales, mainly according to the amount of fresh water coming downstream. Superimposed on these variations, estuarine salinity also fluctuates in relation to the tidal movement of water.

Below, Poole Harbour salinity records from a number of published sources are reviewed, while original unpublished data, made available by the Environment Agency, are also reported. These sources have used various routine methods for the determination of salinity involving the measurement of a parameter relating to salinity such as chloride content or conductivity, and indeed, recent technical definitions of salinity reflect the method of determining it – nowadays conductivity in particular. However, since our present focus is on gross salinity variations from < 1 to > 34 across time and space, variations in the methods of determination in the source publications need not concern us. Therefore, converting from original sources as necessary, here salinity is expressed on the international Practical Salinity Scale (UNESCO, 1981), in which for standard sea water in defined conditions, $S = 35$. While the Practical Salinity Scale has no units, the salinity data as reported here can be considered to correspond numerically with reasonable approximation to a measure of grams per kilogram.

Surface salinity

There is a surface water salinity gradient in Poole Harbour from its mouth to its upper reaches from higher to lower salinity, respectively. Table 1 shows this gradient for seven sites, whose position is illustrated in Figure 2. This information is derived from an extensive data-set established by the Environment Agency using corresponding field and laboratory determinations spread over many years and at all states of the tide.

The relative consistency of salinity maxima across all sites from the harbour mouth deep into the Wareham Channel indicates the extent to which water of marine origin can, especially at high water of spring tides in the summer season, predominate. In contrast, salinity minima decrease much more markedly upstream from the harbour mouth. This phenomenon relates to the fact that salinity minima generally occur at low water when river flow makes up a greater proportion of a water mass much reduced in volume by the ebb tide. Nevertheless, the conspicuous increase in salinity range (maximum – minimum) with progress upstream represents an environmental gradient which will limit the penetration of organisms of both marine and freshwater origin. Between such limits, classic estuarine conditions will exclude all but the genuinely estuarine (euryhaline) species adapted as they are to cope with such a volatile environment.

Salinity variation within the depth

Sea water, with its load of dissolved salt, is denser than fresh water which will, consequently, tend to float above it in the water column. The degree of mixing between

Table 1 Poole Harbour surface salinity characteristics

Site	Maximum salinity	Minimum salinity	Range	Median
1. Harbour entrance	34.7	24.9	9.8	29.8
2a. North Channel	34.5	26.3	8.2	30.4
2b. South Deep	34.1	23.2	10.9	28.65
3a. Poole Bridge	33.9	19.2	14.7	26.55
3b. Hutchins Buoy	34.2	14.2	20.0	24.2
4. Lytchett Bay entrance	30.7	10.6	20.1	20.65
5. Wareham Channel	30.4	1.0	29.4	15.7

Source: Environment Agency.

distinct fresh and salt water masses entering an estuary depends on such factors as the relative volume of river and tidal flows, depth and wind-induced turbulence. Highly stratified estuaries with little mixing exist where large rivers flow into seas with low tidal amplitude. Conversely, where tidal flow dominates, the water column is normally fully mixed and the estuary is classified as vertically homogeneous, i.e. with no variation of salinity with depth. These two conditions represent the ends of a continuum between which lies the partially mixed condition in which, while there is no sudden transition from fresh to salt water, there is nevertheless an increase of salinity with depth (Dyer, 1997).

On the basis of such considerations, Poole Harbour (excluding Holes and Lytchett Bays for which I have been unable to find data) can be considered to consist of two distinct zones: a more or less vertically homogeneous zone in the main body of the harbour, and a partially mixed zone in the Wareham Channel. A July salinity profile as reported by Green (1940) is shown as Figure 3. The partially mixed condition of the Wareham Channel is explicable on the basis that it is through this channel that the main freshwater river flow into the harbour occurs. This, combined with the distance from the harbour mouth, indicates that the freshwater inflow is at its highest in the harbour relative to tidal flows. Conversely, in the main body of the harbour where tidal flows dominate, greater mixing produces a more or less homogeneous salinity profile.

More recent work supervised by Antony Jensen at Southampton University (Barkas, 2001), while confirming this general pattern, provides further information from a series of readings reported over the period April to August 2001. This work shows the extent to which salinity gradients can vary at any one place and includes April data for the Wareham Channel showing a salinity differential from surface to bottom in excess of 10.

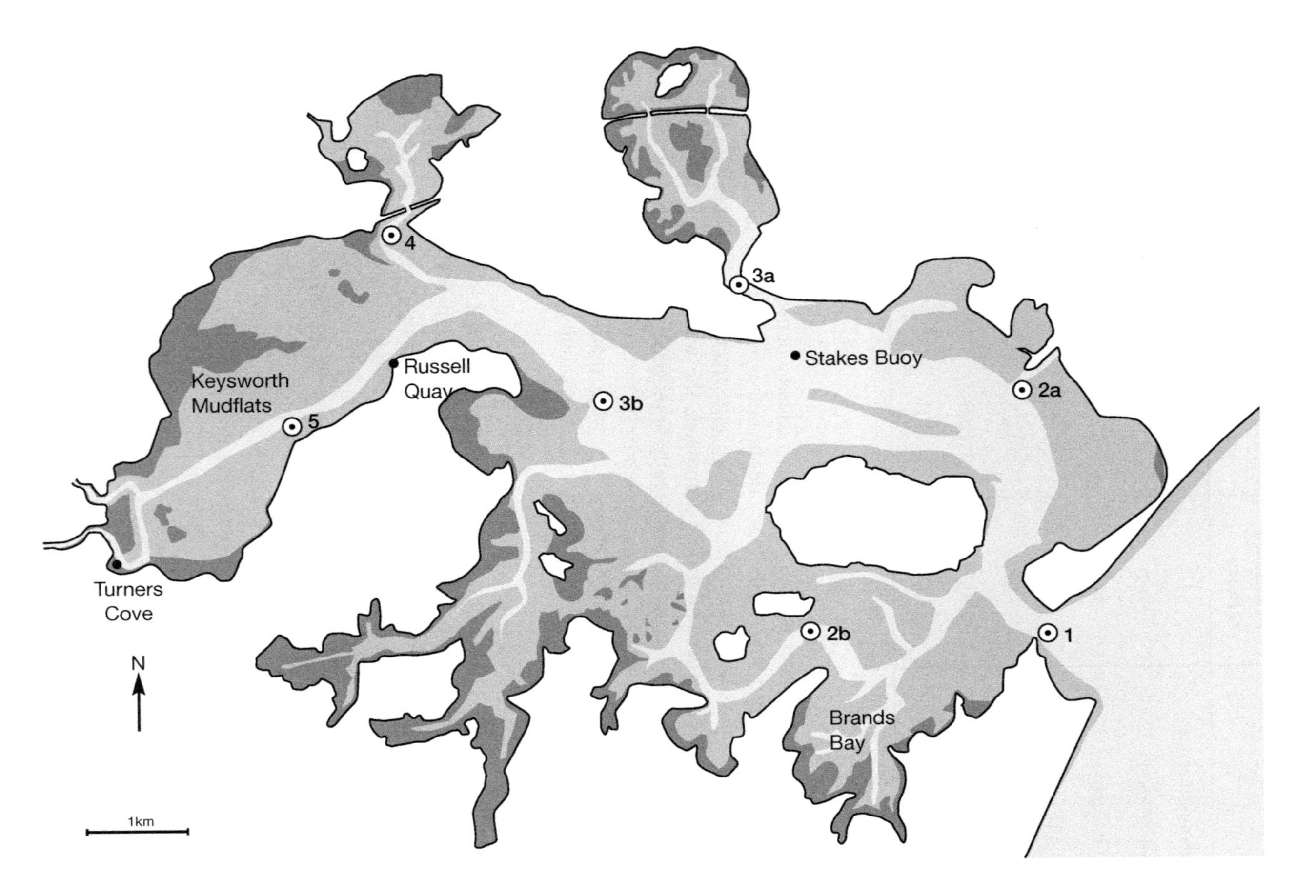

Figure 2 Positions of sampling sites for Tables 1 and 2 and Figure 3.

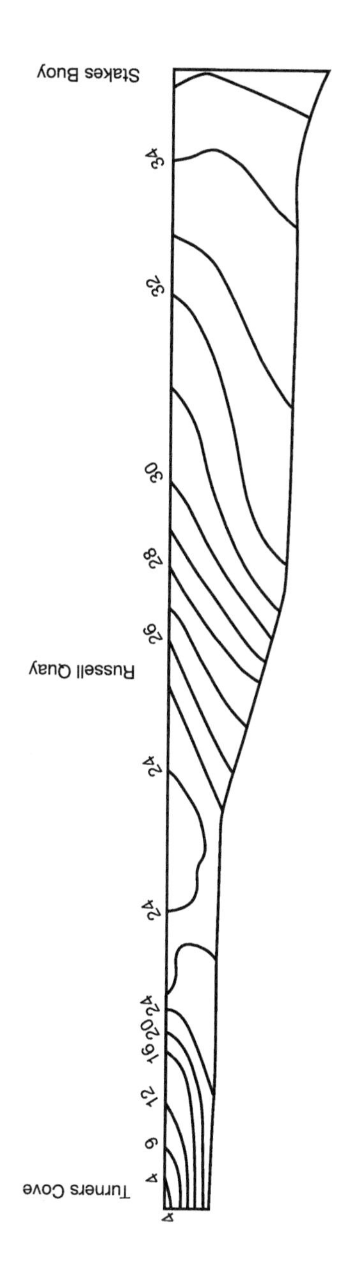

Figure 3 An example of salinity variation with depth from Stakes Buoy along the Wareham Channel (see Figure 2 for site locations).

Interstitial salinity

For many of the animals in Poole Harbour, the ambient salinity (at least at low tide) is that of the interstitial water between the particles of the sediment within which they live. The relationship between the salinity of interstitial water and the water overlying the sediment is known to be dependent on particle size and the slope of the shore. Fine sediments on mudflats are widespread in Poole Harbour and are particularly effective at moderating salinity fluctuations occurring in the overlying water. Through phenomena related to capillary action, fine deposits will tend to retain water and also attenuate evaporation to the air when exposed at low tide (Webb, 1958). This effect combined with the fact that intertidal mudflats are only covered by the higher salinity flood tide, can result in interstitial water of higher mean and lower salinity range than overlying waters (McClusky, 1971). In Poole Harbour, however, factors other than tidal submergence can determine interstitial salinity.

Data from Ranwell *et al.* (1964) shown in Table 2 indicate a general trend of decreased interstitial salinity with increased distance from the harbour mouth – a feature which confirms the importance of tidal submergence. However, superimposed on this pattern are local factors. Measurements on the Arne Bay *Spartina* marshes over a 1 year period showed a seasonal pattern linked to rainfall and connected to the discovery of a water table 20 cm below the marsh surface. This, combined with groundwater seepage, provides the sediment with a steady freshwater supply. In such circumstances as Smith (1955) first showed, the moderating effect of the fine sediment can work to maintain a lower salinity than the prevailing flood tide water.

The ecological effects of such phenomena relate in particular to the distribution of organisms for whom interstitial salinity is important, notably invertebrate infauna and marsh plants. Ranwell *et al.* (1964) for instance, were able to show how, in Poole Harbour, interstitial salinity limited the invasion of *Spartina* marshes by the reed

Table 2 Comparison of interstitial and surface water salinity in *Spartina* marsh at three sites in Poole Harbour

	Straight line distance from harbour mouth (km)	Interstitial water	Surface water	Difference	Date
Brands Bay	2.4	31.0	29.4	+1.6	19.7.62
Arne	6.4	26.7	24.4	+2.3	14.10.62
		25.2	26.4	-1.2	19.7.62
Keysworth	9.3	17.6	20.1	-2.5	19.7.62

Adapted from Ranwell *et al.* (1964).

Phragmites communis and Sea Club Rush *Scirpus maritimus*. However, this work on the saltmarshes is the only information I have been able to find on Poole Harbour interstitial water salinity and there seems to be no published data available on the mudflats despite their significance in terms of invertebrate macrofauna and consequent value to important wader populations and local fisheries. Nevertheless, patterns of invertebrate distribution reported by Caldow *et al.* (this volume, chapter 7) are suggestive of salinity as a limiting factor for various species in Poole Harbour.

Poole Harbour: Estuary or lagoon?

Scientific definitions of an estuary are numerous and varied, and it is not intended to review them here. Fortunately for our purposes, however, there are certain recurring defining attributes in the literature which represent a degree of consensus. These attributes characterize an estuary as: (i) a body of water of mixed origin, part from a discharging river system and part of an adjacent sea which is; (ii) partially enclosed by land, but with; (iii) free connection to the sea and consequently; (iv) subject to the tidal cycles of that sea (e.g. Pritchard, 1967; Barnes, 1974; Fairbridge, 1980). These features collectively have the additional merit of a good alignment with the common understanding of the word.

This chapter so far has examined Poole Harbour in estuarine terms applying, for example, Dyer's (1997) classification of estuaries on the basis of vertical salinity profiles. However, Poole Harbour has also been categorized as a coastal lagoon (Barnes, 1980, 1989, 2000). Insofar as estuaries and lagoons are, in essence, distinct categories of coastal feature – and arguably to be useful scientific concepts they must be – it is worth briefly commenting on which best applies to Poole Harbour.

The literature on coastal lagoons suggests that the concept has proved difficult to define. As for the term estuary, various definitions have been articulated, physical manifestations of which range from relatively small pools to the whole of the Baltic Sea (Jansson, 1981). In any event, coastal lagoons like estuaries are commonly referred to as partially or semi-enclosed bodies of water connected to the sea (e.g. UNESCO, 1980). While they may contain water of mixed origin, this is not a defining attribute as it is with estuaries. Arguably the definitive distinction between the two lies in the nature and extent of the connection with the adjacent sea. While estuaries are partially enclosed they have, by definition, free connection to the sea. Conversely, for lagoons, this connection is restricted with the consequence that the exchange of water between a lagoon and its adjacent sea is attenuated. While such exchange may occur only by percolation through a complete sediment barrier, lagoons may alternatively have one or more entrance channels. However, such channels are sufficiently small in relation to the size of the lagoon that its water body is effectively 'semi-isolated' (Barnes 1980).

Clearly then, in order to decide whether Poole Harbour should be considered a lagoon, we must make a judgement on whether its waters are sufficiently isolated to merit that status. Whilst we know that Poole Harbour has a narrow channel connecting it to the adjacent sea, this does not necessarily mean that the waters of Poole Harbour are to any significant degree isolated from that sea. If the entrance to the harbour, whatever its shape is sufficient to provide free connection to the sea, it should more accurately be considered an estuary. Since these various coastal features represent positions on a continuum, it is understandable that different authorities make different judgements on this, especially if these judgements are based primarily on geomorphology (narrow entrance occluded by sandy spits), rather than more ecologically significant attributes like the degree of isolation of the water mass. In the case of Poole Harbour, it is the latter as evidenced by tidal flows and the salinity regime that resolves the issue.

On tidal coasts, semi-isolation of a lagoonal water mass is taken to mean that much of the water is retained within the lagoon at low tide in the adjacent sea (McClusky and Elliot, 2004). For a lagoon with a river inflow, like an estuary, there will be a longitudinal salinity gradient from high at the seaward end to low at river end. However, in a lagoon, except in the vicinity of the entrance, this gradient is stable in the short term, that is to say it does not fluctuate on a semi-diurnal basis with the tide (Barnes, 1980). This feature contrasts with estuaries within which, due to their open connection to the sea, salinity fluctuates with the ebb and flow of the tide.

Now consider Poole Harbour: does the longitudinal salinity gradient fluctuate with the tide suggesting the 'free connection' that characterizes an estuary or is it stable suggesting a lagoon? Figure 4 shows continuous salinity readings over a 24 hour period juxtapositioned with predicted tide level over the same period. These graphs show that, as far up as the Wareham Channel, salinity varies on a diurnal time-scale which closely matches the tidal cycle.

This feature of Poole Harbour indicates that its entrance while narrow, nevertheless functions as a relatively free connection to the adjacent sea. In this respect, it is worth noting from Admiralty charts, the depth and current speeds of the channel at its narrowest point which, at over 18 m (compared with an average depth for Poole Bay of around 12 m (Riley, 1995)) and up to 2.6 m second^{-1} (Bird and Ranwell, 1964) on the ebb, provide an explanation as to why such a narrow channel does not significantly isolate the harbour's waters from the waters of Poole Bay. On this basis, there appears to be no compelling argument for characterizing Poole Harbour as a lagoon. Conversely we can attribute to it all four of the defining attributes, as outlined above, of an estuary.

If, as I have argued, Poole Harbour is best regarded as an estuary, it is worth reflecting on why it has sometimes been regarded as a lagoon (Barnes, 1980, 1989, 2000) or less definitively as 'lagoon-like' (Gray, 1985), 'lagoonal in character' (English Nature, 1994) or resembling a lagoon (Bird and Ranwell, 1964). The geomorphology of the harbour with its two opposing spits forming a narrow entrance are certainly reminiscent of genuine lagoons but this, as we have seen, is misleading. More relevant to this question is the water level which in Poole Harbour is, as we have observed, above mean tide level

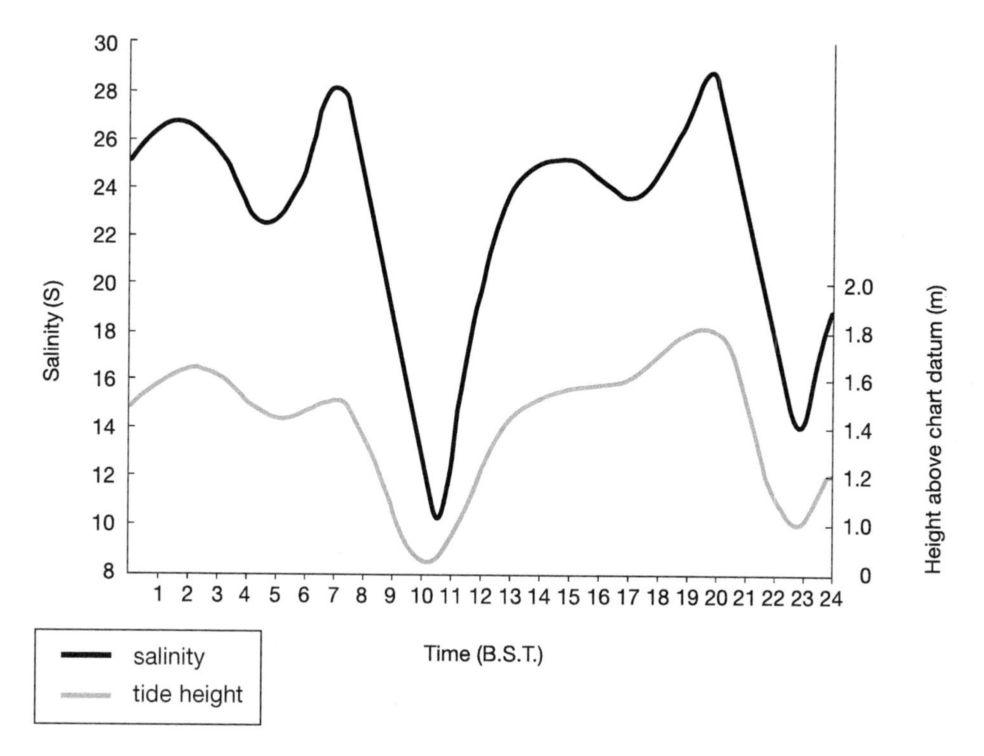

Figure 4 Surface salinity (S) in the Wareham Channel (site 5, Figure 1) over a 24 hour period (24 July 2000) shown in relation to predicted tidal level at Poole Bridge. Sources: Environment Agency; Poole Harbour Tidal Planner 2000.

for about 16 hours a day. However, while this also is reminiscent of lagoons, its cause is classically estuarine. Rather than being due to the retention of water as in a lagoon, the prolonged high is, in contrast, transmitted into the harbour from the adjacent sea.

Finally, it is worth noting that depressions in the sediment surface within the harbour can retain water at low tide to form shallow pools. An example of this is Holton Mere which is known to have persisted for 150 years in the mudflats on the north-western shore of the Wareham Channel disappearing in the first quarter of the twentieth century (although it remains marked on Admiralty charts). This feature, which at one time covered an area of 40 ha, has been described as a transitory shallow lagoon by Hubbard and Stebbings (1968). Arguably, there is a better case for considering such features as small coastal lagoons than for the harbour as a whole, since the water retained in them from a preceding high tide would normally be of relatively high salinity compared with the low tide salinity of the adjacent main channels (although there appears to be no data to support this assertion). Consequently, short-term salinity variations could well be moderated somewhat. However, such pools cannot be considered in any real sense semi-isolated and in any event their existence cannot be persuasively used to argue that the harbour as a whole is a lagoon.

Therefore, while the double high tide effect combined with a microtidal regime can reasonably be described as providing a somewhat lagoon-like environment, such references are misleading, at least insofar as they imply genuine lagoonal characteristics for the harbour. Rather, it is accurate to characterize Poole Harbour as an estuary, and in doing so recognize that estuarine phenomena prevail and provide sufficient explanation for its nature and ecology. With its great size, narrow entrance, double high water, small tidal range and other features of considerable interest, Poole Harbour may be a unique feature of the UK coastline, but it is nonetheless an essentially estuarine feature.

Acknowledgements

Thanks to the Environment Agency (including in particular Danielle Aveling) for making available their extensive Poole Harbour salinity data sets and to Neil Gladwell and Crest Publications for supplying and granting permission to use tidal information from the 'Poole Harbour Planner' for 2000. Thanks also to Antony Jenson and Richard Caldow for critically reading the manuscript. Opinions and errors remain the responsibility of the author.

References

Barkas, N. (2001) Summer Plankton Community Survey in Poole Harbour, Dorset, with Special Reference to the Role of Dissolved Organic Carbon. MSc. dissertation. Southampton: School of Ocean and Earth Science, University of Southampton.

Barnes, R. S. K. (1974) *Estuarine Biology*. London: Arnold.

Barnes, R. S. K. (1980) *Coastal Lagoons*. Cambridge: Cambridge University Press.

Barnes, R. S. K. (1989) The coastal lagoons of Britain: an overview and conservation appraisal. *Biological Conservation*, **49**: 295–313.

Barnes, R. S. K. (2000) The Fleet, Dorset, in relation to other coastal lagoons. In: *The Fleet Lagoon and Chesil Beach*. Revised Edition. Carr, A. P., Seaward, D. R. and Sterling, P. H. (eds). Dorchester: Dorset County Council and the Chesil Beach and the Fleet Nature Reserve.

Bird, E. C. F. and Ranwell, D. S. (1964) *Spartina* marshes in Southern England iv. The physiography of Poole Harbour, Dorset. *Journal of Ecology*, **52**: 355–366.

Dyer, K. R. (1997) *Estuaries: A Physical Introduction*. Second Edition. Chichester: John Wiley.

English Nature (1994) *Important Areas of Marine Wildlife Around England*. Peterborough: English Nature.

Fairbridge, R. (1980) The estuary: its definition and geodynamic cycle. pp. 1–35. In: *Chemistry and Geochemistry in Estuaries*. Olausson, E. and Cato, I. (eds). New York: John Wiley.

Falconer, R. A. (1984) Temperature distributions in a tidal flow field. *Journal of Environmental Engineering*, **110**: 1099–1116.

Gray, A. J. (1985) *Poole Harbour: Ecological Sensitivity Analysis of the Shoreline*. Abbots Ripton: NERC Institute of Terrestrial Ecology.

Green, F. H. W. (1940) *Poole Harbour, A Hydrographic Survey 1938–9*. London: Geographical Publication Ltd for Poole Harbour Commissioners and University College Southampton.

Hubbard, J. C. E. and Stebbings, R. E. (1968) *Spartina* marshes in Southern England vii. Stratigraphy of the Keysworth Marsh, Poole Harbour. *Journal of Ecology*, **56**: 702–722.

Jansson, B.-O. (1981) Production dynamics of a temperate sea – the Baltic. In: *Coastal Lagoon Research, Present and Future. UNESCO Technical Papers in Marine Science,* No. 33. Paris: UNESCO.

McClusky, D. S. (1971) *Ecology of Estuaries*. London: Heinemann.

McClusky, D. S. and Elliot, M. (2004) *The Estuarine Ecosystem*. Third Edition. Oxford: Oxford University Press.

Pritchard, D. W. (1960) Estuaries. In: *Encyclopaedia of Science and Technology*. New York: McGraw-Hill.

Pritchard, D. W. (1967) What is an estuary: a physical viewpoint. pp. 3–5. In: *Estuaries*. Publication No. 83. Washington DC: American Association for the Advancement of Science.

Ranwell, D. S., Bird, E. C. F., Hubbard, J. C. E. and Stebbings, R. E. (1964) *Spartina* salt marshes in Southern England V. Tidal submergence and chlorinity in Poole Harbour. *Journal of Ecology*, **52**: 627–641.

Riley, M. J. (1995) An investigation into apparent anomalies of the tidal behaviour in Poole and Christchurch Bays, UK. *The Hydrographic Journal,* **76**: 29–33.

Smith, R. I. (1955) Salinity variation in interstitial water at Kames Bay, Millport, with reference to the distribution of *Nereis diversicolor. Journal of the Marine Biological Association of the UK,* **34**: 33–36.

Tait, R. V. (1981) *Elements of Marine Ecology*. Third Edition. London: Butterworths.

UNESCO (1980) *Coastal Lagoon Survey. UNESCO Technical Papers in Marine Science,* No. 31. Paris: UNESCO.

UNESCO (1981) *Tenth Report for the Joint Panel on Oceanographic Tables and Standards. UNESCO Technical Papers in Marine Science*, No. 32. Paris: UNESCO.

Webb, J. E. (1958) The ecology of Lagos Lagoon V. Some physical properties of lagoon deposits. *Philosophical Transactions of the Royal Society B*, **241**: 393–419.

4. The Vegetation of Poole Harbour

Bryan Edwards

Dorset Environmental Records Centre, Library Headquarters, Colliton Park, Dorchester, Dorset DT1 1XJ

The vegetation of the 423 ha of saltmarshes present within Poole Harbour is described. The formation of the saltmarsh is linked to the complex tidal regime and to the invasion of *Spartina anglica*. Three saltmarsh zones, lower, middle and upper, are defined and within each zone the different plant communities are described in the context of the National Vegetation Classification. Communities found on sand, shingle and in transitional zones to terrestrial habitats are also briefly described and are of ecological importance. A summary of the rare and scarce plant species present within the harbour is provided.

Introduction

This chapter covers all intertidal vegetation within the harbour, up to and including those areas inundated at the highest spring tides. Sand and shingle habitats are also included as are transitions from saltmarsh to other terrestrial habitats. The major reedbeds were not surveyed as they were the subject of a specific study in 2000 (Cook, 2001), but the smaller stands not covered in that report are included. The chapter is a summary of fieldwork undertaken during the summers of 2001 and 2002 (Edwards, 2004).

For the purposes of this chapter, all plant names are in Latin only, and follow Stace (1991). Nomenclature for plant communities follows the National Vegetation Classification (NVC) (Rodwell,1995, 2000).

Vegetation

The vegetation of the harbour is strongly linked to the complex tidal regime, with a double high tide and a small tidal range of 1.8 m at spring tides and 60 cm at neaps (Hubbard and Stebbings, 1968). These factors coupled with the narrowness of the entrance produce a lagoon-like effect, with relatively poor flushing capabilities. During spring tides, substantial parts of saltmarsh are inundated for long periods. Another important factor in the formation of saltmarsh is the colonization and then rapid expansion of *Spartina anglica* (Raybould, 1997). Now the steady decline of *S. anglica* coupled with the predicted rise in sea levels will doubtless shape the saltmarshes of the future. Saltmarsh vegetation by its very nature is species-poor, typically being dominated by one or two highly specialized halophytic species. Generally, vegetation occurs in distinct zones linked to the tidal regime of a particular site (Tables 1 and 2).

The tidal range in Poole Harbour is very small, resulting in relatively poor zonation of the vegetation.

Summary of areas of habitats

Approximately 423 ha of saltmarsh (Table 3) are found at present within the harbour. The vast majority of this is lower saltmarsh dominated by *Spartina anglica* and is frequently inundated. Middle and upper vegetation is limited in extent and frequently found as linear stands just below mean high water.

These broad habitat categories have been split further into habitats listed in Annex I of the EU Habitats Directive (European Commission, 1996) (Table 4). The UK has a responsibility to protect these habitats.

Plant communities

The present survey was not intended to be a full NVC survey of the harbour, but 152 quadrats were made throughout the harbour, using the standard NVC methodology, to sample the diversity of the plant communities within the site.

A total of 39 NVC communities, sub-communities and variants were identified, along with five vegetation types not described in the NVC (Table 5).

Table 1 Definition of saltmarsh zones

Lower saltmarsh
This zone extends from low water mark through pioneer vegetation of *Salicornia* spp. and *Spartina anglica*, to closed, stable communities of *Aster tripolium*, *Limonium vulgare*, *Plantago maritima*, *Puccinellia maritima* and *Triglochin maritimum*.

Middle saltmarsh
A difficult zone to distinguish in the harbour due to the small tidal range and subsequent compact nature of many of the marshes. Species absent from the lower zone such as *Festuca rubra*, *Juncus gerardii* and *J. maritimus* can dominate here but also extend into the upper zone.

Upper saltmarsh
Agrostis stolonifera, *Elytrigia atherica*, *Festuca rubra*, *Juncus gerardii* or *J. maritimus* typically dominate this zone. Species tolerant of both freshwater and slightly saline conditions such as *Eleocharis uniglumis*, *Juncus subnodulosus*, *Potentilla anserina*, *Samolus valerandi* and *Triglochin palustris* occur.

Table 2 Approximate zonation of the saltmarsh communities within Poole Harbour

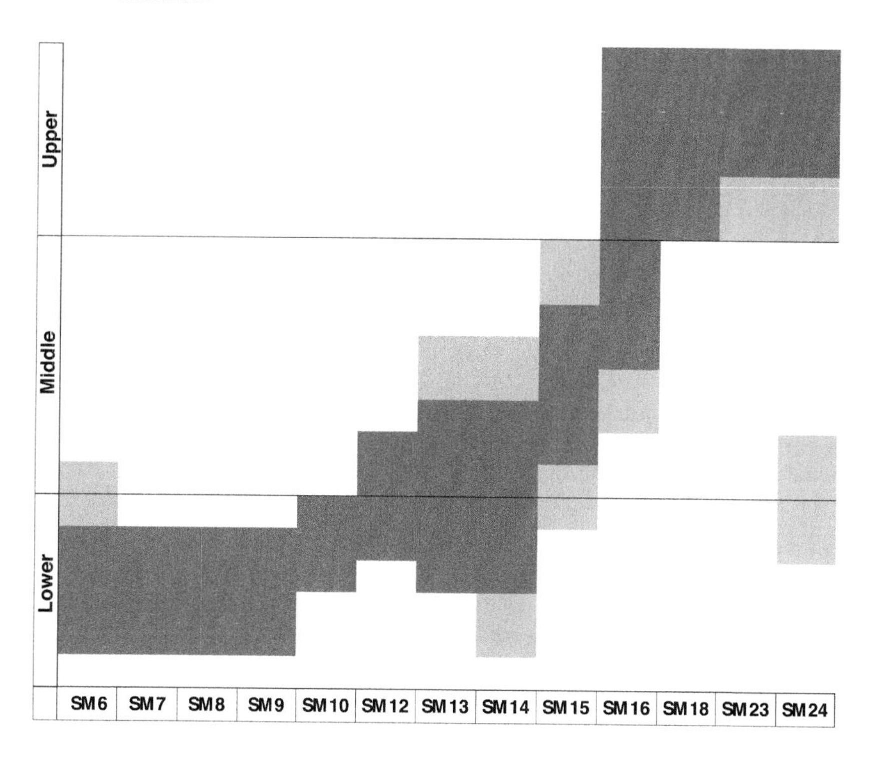

SM6	*Spartina anglica* saltmarsh
SM7	*Anthrocnemum perenne* stands
SM8	Annual *Salicornia* saltmarsh
SM9	*Suaeda maritima* saltmarsh
SM10	*Puccinellia maritima-Salicornia-Suaeda maritima* saltmarsh
SM12	*Aster tripolium* saltmarsh
SM13	*Puccinellia maritima* saltmarsh
SM14	*Halimione portulacoides* saltmarsh
SM15	*Juncus maritimus-Triglochin maritimum* saltmarsh
SM16	*Festuca rubra* saltmarsh
SM18	*Juncus maritimus* saltmarsh
SM23	*Spergularia marina-Puccinellia distans* saltmarsh
SM24	*Elymus pycnanthus* saltmarsh

Table 3 Summary of areas of different habitats

Habitat	Area
Saltmarsh total	423 ha
Lower saltmarsh	358 ha
Middle-upper saltmarsh	65 ha
Atlantic salt-meadows	175 ha
Mediterranean salt-meadows	33 ha
Salicornia and other annuals colonizing mud and sand	2.76 ha
Mediterranean and thermo-Atlantic halophilous scrub	0.51 ha
Shifting dunes along the shoreline with *Ammophila arenaria*	0.67 ha
Perennial vegetation of stony banks	0.20 ha
Annual vegetation of driftlines	0.16 ha

Table 4 Definition of Annex I habitats present in Poole Harbour

Salicornia and other annuals colonizing mud and sand
Salicornia spp. and *Suaeda maritima* on bare mud or sand substrates in the lower saltmarsh.

Atlantic salt-meadows
A wide habitat type encompassing NVC communities SM10 to SM20 inclusive. It includes all lower, middle and upper marsh vegetation dominated by one or more of *Agrostis stolonifera*, *Aster tripolium*, *Atriplex portulacoides*, *Eleocharis uniglumis*, *Festuca rubra*, *Juncus gerardii*, *Plantago maritima*, *Puccinellia maritima* and *Triglochin maritimum*.

Mediterranean salt-meadows
Juncus maritimus dominated NVC communities, SM15 and SM18, are included here.

Shifting dunes along the shoreline with *Ammophila arenaria*
Embryonic and mobile dunes dominated by *Ammophila arenaria* are included in this habitat, although *Leymus arenarius* dunes are excluded. SD6 is the corresponding NVC type.

Perennial vegetation of stony banks
Sparsely vegetated stabilized shingle within the NVC community. SD1 is included here.

Annual vegetation of driftlines
The SD2 *Honckenya peploides* strandline community, and vegetation with various *Atriplex* spp. and *Beta vulgaris*, which is not referable to any NVC community, are included in this category.

Mediterranean thermo-Atlantic halophilous scrub
Suaeda vera dominated vegetation with *Atriplex portulacoides* and *Elytrigia atherica*, and stands of *Sarcocornia perennis* are included here. SM25 and SM7 are the corresponding NVC communities.

Lower saltmarsh communities

Lower saltmarsh communities are not particularly diverse within the harbour as *Spartina anglica* dominates this zone. However, local pioneer communities are present as small stands on firm mud or sand substrates. Brownsea Lagoon has the largest stands of

Table 5 Summary of plant communities by habitat within Poole Harbour Special Protection Area (SPA)

Habitat	Number of plant communities
Saltmarsh	25
Sand-dune, shingle and driftline	9
Brackish swamps	7
Brackish grassland	2
Transitional vegetation	1
Total	**44**

pioneer *Salicornia* marsh (SM8) in which *S. ramosissima* and *S. europaea* dominate with the rarer, and nationally scarce, *S. fragilis* present locally. Smaller stands are found scattered along the southern shore. The nationally scarce *Sarcocornia perennis* is locally frequent among *Spartina*, but in the east of the harbour it locally forms dense stands (SM7) on firm sandy substrates. *Suaeda maritima* is generally rarer than the annual *Salicornia* spp., but in three areas it forms distinctive communities (SM9) on firm mud.

The *Spartina anglica* saltmarsh (SM6) is by far the most widespread of all the lower saltmarsh communities. *Spartina anglica* forms dense swards of between 20 cm and 80 cm in height. At the lowest level on soft mud, there are no other associates apart from marine algae. On firmer substrates, there are occasional scattered plants of *Atriplex portulacoides*, *Limonium vulgare*, *Salicornia* spp., *Sarcocornia perennis*, *Spergularia media* and *Suaeda maritima*. In places, a distinctive *Spartina anglica-Puccinellia maritima* saltmarsh (not described in the NVC) is well developed. *Spartina anglica* and *Puccinellia maritima* are co-dominant with few other species, except for *Atriplex portulacoides* and *Sarcocornia perennis*, attaining any abundance. These two communities occupy approximately 320 ha of the lower marsh resource in the harbour.

On firmer substrates, two species-rich, and colourful, communities are well developed. In the west of the harbour *Aster tripolium* is abundant (SM12a) with *Atriplex portulacoides*, *Plantago maritima* and *Puccinellia maritima* all prominent. In the south and east of the harbour, this community is replaced by the distinctive *Limonium vulgare* saltmarsh (SM13c). *Limonium vulgare* forms dense patches and is typically accompanied by *Armeria maritima*, *Atriplex portulacoides*, *Cochlearia anglica*, *Plantago maritima*, *Puccinellia maritima*, *Spergularia media* and *Triglochin maritimum*. In places, heavy grazing has reduced the abundance of herbs and *Puccinellia maritima* dominates (SM13a), with *Salicornia europaea* locally abundant. The increase of Sika Deer grazing is having more of an impact than the limited amount of cattle grazing and needs to be monitored.

A feature of the extensive *Spartina* marshes is the complex network of creeks. During high tides a small amount of sediment is deposited at the sides of the creeks eventually raising them slightly. These creek levees often support different vegetation from the adjoining marsh. Particularly characteristic are linear stands of *Atriplex portulacoides* (SM14), which is typically the sole dominant, although scattered plants of *Puccinellia maritima*, *Spartina anglica* and *Sarcocornia perennis* may be present.

Middle saltmarsh communities

This zone is very difficult to distinguish in Poole Harbour due to the complex tidal regime. However, in places the *Juncus maritimus-Triglochin maritimum* saltmarsh (SM15) occupies this zone. *Juncus maritimus* is the overwhelming dominant forming dense clonal stands with *Atriplex portulacoides* and *Limonium vulgare* both locally abundant. *Juncus gerardii* and *Plantago maritima* are both prominent in some stands, but generally associated species are few. Stands of this community are often small and linear.

Upper saltmarsh communities

A total of 65 ha of upper saltmarsh is present in the harbour. Much of it is in narrow or linear stands at high water mark. There are two main communities, both of which are species-rich, compared with the lower saltmarsh communities, and of high ecological interest.

Most widespread is a short community dominated by *Festuca rubra* and *Juncus gerardii* (SM16). *Armeria maritima* and *Glaux maritima* are abundant locally (SM16c), with *Limonium vulgare*, *Plantago maritima*, *Puccinellia maritima*, *Spergularia media* and *Triglochin maritimum* also present. Where there is freshwater seepage into the back of the marsh, a different range of associates are present with *Agrostis stolonifera*, *Leontodon autumnalis* and *Potentilla anserina* particularly characteristic (SM16e).

The other major community is largely confined to the southern shore between Arne and Bramble Bush Bay. The *Juncus maritimus* saltmarsh (SM18b) is easily distinguished by the abundance of *J. maritimus* tussocks, between which *Agrostis stolonifera*, *Festuca rubra* and *Juncus gerardii* form a grassy sward. The scattered plants of *Oenanthe lachenalii* give the vegetation a distinctive appearance in late summer. An interesting variant is found on the Arne peninsula in a transition from saltmarsh to heathland where *Schoenus nigricans* is co-dominant with *Juncus maritimus*. This type of vegetation is not known elsewhere in southern England.

The only other community commonly encountered in this zone is the *Elymus atherica* saltmarsh (SM24). *Elymus atherica* forms dense stands up to 1 m high on firm substrates at the back of the saltmarsh or more rarely along creek levees. Often no other species attain any abundance, but there may be scattered plants of *Atriplex patula*, *A. prostrata*, *Rumex crispus* and *Sonchus arvensis*.

Sand-dune, shingle and driftline communities

Scattered throughout the harbour are small areas of shingle and sand. Where tourist pressure is not too great, a sparse covering of vegetation may develop. The communities present are often small in extent and support species that are otherwise rare in the harbour, and are under more threat than those found on the saltmarshes.

Small areas of vegetated shingle, often mixed with sand, occur at four places. They are sparsely vegetated, but support scattered plants of *Rumex crispus* and *Silene uniflora* (SD1), plus *Atriplex prostrata*, *Beta vulgaris* ssp. *maritima* and *Tripleurospermum maritimum*. More locally *Atriplex glabriuscula*, *A. littoralis*, *Elymus atherica*, *Senecio jacobaea*, *S. sylvaticus* and *S. viscosus* are also present.

At three sites, *Honckenya peploides* is locally abundant and forms a distinctive strandline community (SD2). Prostrate mats of *Honckenya peploides* dominate with no other species attaining any abundance. Associated species may include other uncommon strandline species including *Atriplex laciniata*, *A. littoralis* and *Cakile maritima*.

Sandy beaches and small dunes have a different flora. *Ammophila arenaria* and *Carex arenaria* are the most abundant species. Stands of the robust tussock-forming *Ammophila arenaria* (SD6) are characteristic of young mobile dunes in the east of the harbour. Associated species are few, apart from scattered plants of *Atriplex prostrata*, *Elytrigia atherica*, *Hypochaeris radicata*, *Leymus arenarius*, *Senecio jacobaea*, *S. sylvaticus* and *S. viscosus*. Locally the equally robust *Leymus arenarius* replaces *Ammophila arenaria* and forms a distinctive community (SD5), with few associates except for occasional plants of *Ammophila arenaria*, *Atriplex prostrata*, *Elytrigia atherica* and uncommonly, *E. juncea*.

Carex arenaria is widespread around the harbour wherever there are sandy substrates. However, only locally does it become dominant enough to form a distinct community (SD10). *Carex arenaria* is the sole dominant with no other species attaining any abundance. Associated species include species typical of dry acid grasslands such as *Aira praecox*, *Brachythecium albicans*, *Erodium cicutarium*, *Hypochaeris radicata*, *Jasione montana* and *Senecio jacobaea*.

At the junction of the sand, shingle and saltmarsh, a distinctive and important community is present in five sites. The nationally scarce woody shrub *Suaeda vera* dominates forming dense stands up to 1.5 m tall. Between the bushes, a low sward of *Atriplex portulacoides* and *Elytrigia atherica* is usually present. This type of Mediterranean-Atlantic scrub is a rare habitat within the UK.

Brackish swamp communities

Stands of tall graminoids are a feature of the western part of the harbour, often where there is some freshwater seepage. The most familiar are the brackish reedbeds which are a prominent and important feature of the harbour. A recent survey of the larger reedbeds (Cook, 2001) found there to be 102.33 ha of reedbed split between 12 sites, with Holton Heath, Lytchett Bay and Slepe Moor supporting the largest areas. Botanically the reedbeds (S4) are species-poor, with dense *Phragmites australis* often dominating to the exclusion of all other species. Where there is freshwater seepage at the back of the reedbeds, poor-fen species such as *Galium palustre, Hydrocotyle vulgaris, Juncus subnodulosus, Lotus pedunculatus, Mentha aquatica* and *Oenanthe crocata* may be present. Grazed stands tend to be slightly richer with a grassy layer of *Agrostis stolonifera, Festuca rubra, Juncus gerardii* or *Spartina anglica* beneath the robust *Phragmites*.

Bolboschoenus maritimus is a common species around the harbour, typically at the back of saltmarshes, and extending locally to brackish ditches in grazing marsh. Stands of *Bolboschoenus maritimus* (S21) are common with *B. maritimus* the overwhelming dominant forming dense monotonous stands up to 2 m high. Typically there are few other species present apart from scattered plants of *Atriplex portulacoides, Schoenoplectus tabernaemontani* and *Spartina anglica*. More locally the glaucous *Schoenoplectus tabernaemontani* forms small stands (S20). Where there is freshwater flow into the back of the saltmarsh, species-rich vegetation develops with *Agrostis stolonifera, Eleocharis uniglumis, Festuca rubra, Juncus gerardii* and *Oenanthe lachenalii*, plus poor-fen species such as *Galium palustre, Hydrocotyle vulgaris* and *Lycopus europaeus*.

Transitional vegetation

One of the features of the southern shore of the harbour is the stands of transitional vegetation. Of particular interest are those from mire to saltmarsh, which are possibly unique to the area. During the survey three different types of transition were noted.

(i) Mire-saltmarsh

A rare transition, possibly unique to the harbour in southern England, largely confined to three sites along the southern shore, and which typically comprises dense stands of *Juncus subnodulosus*, sometimes with scattered plants of *Schoenus nigricans* and *Molinia caerulea*.

(ii) Freshwater seepage

Juncus subnodulosus is also abundant in this type, but is joined in some sites by *Eleocharis uniglumis, Isolepis cernua, Samolus valerandi, Schoenoplectus tabernaemontani* and *Triglochin palustre*.

(iii) Grassland-saltmarsh

This type is characterized by dense swards of *Agrostis stolonifera*, often with *Alopecurus geniculatus, Festuca rubra* and *Potentilla anserina*. Notable plants species include *Carex distans, Lotus glaber* and *Trifolium fragiferum*.

Transitional communities

Brackish grassland is largely confined to the grazing marsh in the lower floodplains of the Rivers Frome, Piddle and Sherford, but locally there are small stands in fields adjacent to the upper saltmarsh around the harbour. *Agrostis stolonifera* is the dominant grass species (MG11), with *Alopecurus geniculatus*, *Festuca rubra*, *Holcus lanatus* and *Lolium perenne* all locally prominent. *Carex distans*, *C. otrubae* and *Juncus gerardii* may be present in the richer stands. Herbs are generally restricted to frequent *Trifolium repens* and *Potentilla anserina*. Formerly the nationally scarce *Alopecurus bulbosus* was a feature of these grasslands, but a reduction in salinity with the construction of seawalls has reduced it to three small populations, its place being taken by the hybrid *A. plettkei*.

Locally within these grasslands small brackish pans develop where water stands at high spring tides. In these pans a distinctive community (SM23) is present and dominated by small prostrate plants of *Spergularia marina*. There is a wide range of associates including *Puccinellia maritima*, *Atriplex prostrata*, *Gnaphalium uliginosum*, *Juncus bufonius*, *J. gerardii*, *Plantago major* and *Polygonum aviculare*.

Very different vegetation develops on peaty soils where there is freshwater seepage into the back of saltmarshes. It is not sufficiently described within the NVC, although some stands clearly have affinities to the M22 *Juncus subnodulosus-Cirsium dissectum* fen-meadow, and could be referred to as the *Oenanthe lachenalii* sub-community (non-NVC). *Juncus subnodulosus* dominates, typically forming dense stands up to 1.5 m high. Beneath the *Juncus* there is often frequent *Agrostis stolonifera* and *Festuca rubra*, or more rarely *Eleocharis uniglumis*. *Juncus maritimus* is locally prominent. The herb component is poor and largely confined to scattered plants of *Oenanthe lachenalii* and *Galium palustre*, with occasional *Leontodon autumnalis*, *Lotus pedunculatus* and *Samolus valerandi*. In similar situations, stands of shorter vegetation dominated by *Eleocharis uniglumis* (SM20) are found in three areas along the southern shore of the harbour where there is freshwater seepage into the back of the marsh. *Eleocharis uniglumis* is accompanied by *Agrostis stolonifera*, *Festuca rubra* and *Juncus gerardii* giving a ‘grass-dominated’ appearance to the community.

Rare, scarce and notable species

Few rare or scarce plant species are found within the true saltmarsh habitats within the harbour. At present two Red Data Book (RDB) and ten Nationally Scarce (NS) species are recorded from the harbour (Table 6), but only four of these are strongly associated with saltmarsh habitats.

Along with the above nationally important species, there are a number of species that are rare or uncommon within the county or almost confined to the harbour (Table 7). Only four of these are found in the saltmarshes. Importantly, three species are found on strandlines in the east of the harbour and are under immediate threat from pressures of tourism.

Table 6 Rare and scarce plant species within Poole Harbour

Species	Status	Preferred habitat
Alopecurus bulbosus	NS	Brackish grazing marsh
Atriplex longipes	NS	Upper saltmarsh
Carex punctata	NS	Brackish grassland
*Cynodon dactylon**	RDB-Vu	Stabilized sand-dunes
Festuca arenaria	NS	Stabilized sand-dunes
Polypogon monspeliensis	NS	Brackish grazing marsh
Ruppia cirrhosa	NS	Brackish water
Salicornia fragilis	NS	Lower saltmarsh
Salicornia pusilla	NS	Lower saltmarsh
Sarcocornia perennis	NS	Lower saltmarsh
*Scirpoides holoschoenus**	RDB-Vu	Stabilized sand-dunes
Suaeda vera	NS	Strandlines

*Not considered native in Dorset.

Table 7 Species of local interest within Poole Harbour

Species	Preferred habitat
Ammophila arenaria[3]	Sand-dunes
Atriplex laciniata[1]	Sandy strandlines
Carex extensa[3]	Upper saltmarsh
Elytrigia juncea[3]	Sandy strandlines
Hypochaeris glabra[2]	Stabilized sand-dunes
Honckenya peploides[2]	Sandy strandlines
Leymus arenarius[3]	Stabilized sand-dunes
Limonium vulgare[3]	Mid and lower saltmarsh
Puccinellia distans[2]	Upper saltmarsh
Ruppia maritima[2]	Brackish water
Salicornia dolichostachya[2]	Lower saltmarsh

[1] Dorset Rare – fewer than three sites in the county.
[2] Uncommon in the county.
[3] Confined to, or very rare outside Poole Harbour.

Acknowledgements

The survey work forming the basis of this chapter was funded by English Nature and the Environment Agency.

References

Cook, K. (2001) *Poole Harbour Reedbed Year 2000 Survey Summary Report.* Purbeck Reedbed Working Group.

European Commission (1996) *Interpretation Manual of the European Union Habitats* (Version EUR15). Brussels: European Commission.

Raybould, A. F. (1997) The history and ecology of *Spartina anglica* in Poole Harbour. *Proceedings of the Dorset Natural History and Archaeological Society*, **119**: 147–158.

Rodwell, J. S. (ed.) (1995) *British Plant Communities* Volume 4: *Aquatic Communities, Swamps and Tall-herb Fens*. Cambridge: Cambridge University Press

Rodwell, J. S. (ed.) (2000) *British Plant Communities* Volume 5: *Maritime Communities and Vegetation of Open Habitats*. Cambridge: Cambridge University Press.

Stace, C. A. (1991) *New Flora of the British Isles*. Cambridge: Cambridge University Press.

The Ecology of Poole Harbour
John Humphreys and Vincent May (editors)

5. Physical and Ecological Aspects of the Poole Harbour Reedbeds

Kevin Cook

Fieldwork Ecological Services Ltd, Church Farm House, 70 Back Lane,
Okeford Fitzpaine, Blandford Forum, Dorset DT11 0RD
e-mail: kevin@cook5381.freeserve.co.uk

UK reedbeds are a high priority habitat to protect as so many have been lost to agricultural improvements. In 2000, the Purbeck Biodiversity Reedbed Working Group commissioned a survey of the 13 most significant reedbeds within Poole Harbour, to assess their condition and conservation value. It concluded that most of the reedbeds were in good condition supporting a number of notable species. Of the six nationally rare Red Data bird species cited in the *UK Costed Habitat Action Plan for Reedbeds*, Marsh Harrier, Cetti's Warbler and Bearded Tit are regular users of the Purbeck reedbeds. Some problems highlighted were: damage by Sika Deer, scrub encroachment, varying water quality control, limited conservation management, rising sea levels, loss by erosion and habitat change. The survey report proposed a number of possible practical management opportunities and some theoretical areas for reedbed expansion.

Introduction

The reedbed habitat, characterized by the dominance of the common reed *Phragmites australis,* forms on permanently wet or frequently flooded freshwater or tidal land. Reedbeds are noted for their importance to a range of specialized species, several limited solely to reedbeds. Over the last century, many UK reedbeds have been lost to agricultural improvements as their value for thatching declined or as part of larger wetland reclamation schemes. Reedbeds with their specialized wildlife have, therefore, become a high priority habitat to protect.

As relatively little is known about the reedbeds of Dorset, the Purbeck Biodiversity Reedbed Working Group (as part of the Biodiversity Action Plan for reedbeds) commissioned this survey of the 13 most significant reedbeds or groups of reedbeds within Poole Harbour to assess their condition and conservation value. They were surveyed in October/November 2000 (Table 1). For full details, refer to Cook (2001).

Poole Harbour and its reedbeds

Poole Harbour and its margins are internationally important for wildlife. The harbour and most of the reedbeds associated with it are included within European and UK designations designed to protect wildlife. All the reedbeds are Sites of Special Scientific Interest (SSSIs) and most are within the Poole Harbour Special Protection Area (SPA) classified in 1999, and the Ramsar Site boundary.

Estimates of areas of reedbeds are likely to vary as they often have transitional zones with other habitats. However, Dorset's reedbeds occupy at least 285 ha – about 47% of the whole reedbed area of south-west England. The bulk of these (174.55 ha) are within the Poole Harbour environs, representing over 60% of the county's reedbeds and about 30% of the reedbed total for the south-west of England.

Some reedbeds merge into the harbour down the river valleys, e.g. the Moors is an extension of the reedbeds lining the edge of the River Frome, the South Middlebere reedbeds grow alongside the Corfe River and the northern parts of the Holton Lee reedbeds grow alongside the Sherford River. Others are part of the natural zonation of the harbour edge: above the saltmarsh and below fringing carr woodlands and bogs such as at Holton Heath and Lytchett Bay.

Some beds, such as the Moors and Swineham, are on the harbour side of a seawall and are mainly tidal, cut off from significant freshwater influence. Others are inside the seawall or far enough up a valley to have a large freshwater influence: Brownsea, Salterns, Slepe and the Middlebere reedbeds show this. Brownsea reedbed 1 and the South Middlebere reedbeds are the only ones studied which are purely freshwater, the rest being tidal to some degree. Shag Looe and Bucks Cove are outliers, growing on spits in the harbour and surrounded by saltmarsh.

Threats to the Poole Harbour reedbeds

Though after this survey most beds have been classified as stable within the context of a short time-scale (say 5 years), in the long term, many could decline in quality from:

- Increasing deer damage – deer are a particular problem at East Holton but all reedbeds, excluding Swineham Point, had evidence of deer and most had some localized grazing or worse.
- Drying out causing curved and sparse reed growth. Curving also occurs where there is a change from saline to freshwater. Curved reeds often grow partly parallel to the ground being of less value to nesting Reed Warblers and creating a dense mat that can prevent access to the reedbed floor by Snipe and Water Rails.
- Increase of salinity in artificially maintained freshwater beds where water control systems are not functioning well – Salterns and Slepe reedbeds are artificially freshwater, the saline influence being partly held back by a seawall and sluices.

- Rising sea levels – it has been predicted that within the next 80 years there could be a 12–67 cm rise in sea level that would significantly change the area and quality of most of the beds. Erosion at the harbour edge is already evident in western beds, e.g. Swineham Point.
- Scrub encroachment in freshwater beds. Scrub is not a problem in tidal areas and is partly held back by deer elsewhere. The South Middlebere beds have a strong invasion of sallow and bog myrtle.
- Uncontrolled cattle grazing – this is only a local problem in Middlebere (north-west of the stream and north) and Wych Lake east where cattle trample reeds and degrade the habitat.
- Lack of beneficial wildlife monitoring and management.

Poole Harbour reedbed survey summary

Reedbed dossiers

The survey during October and November 2000 provided a set of dossiers in which current information about a bed is centralized in whole or summary form. This includes recent past wildlife records, conservation management and historical information and the survey results. Each reedbed was photographed to record significant features and the photographs are recorded on CD ROM.

Reedbed condition

Reedbed condition was assessed by recording:

(i) straightness, width and height of the reed stems – curved, narrow, short reeds are a negative feature
(ii) number of reed stems per square metre (averaged from several counts per bed) – low reed counts are a negative feature
(iii) hardness/softness of reed stems – soft reeds are a negative feature
(iv) depth of litter and surface water – lack of surface water is a negative feature in freshwater beds
(v) evidence of negative features such as grazing and trampling, scrub encroachment, drying out
(vi) the hydrology – salinity (recorded by a salinity meter), presence or absence of flowing freshwater, open water and ditches
(vii) adjacent land use – adjacent land of low ecological value scored as a negative feature
(viii) current management and management features including sluices and reedbed cutting.

Data were recorded in detail within the dossiers on Reedbed Assessment Forms to enable surveys to be easily repeated in the future. From these forms an assessment of reedbed condition, based on the above factors, has been summarized in Table 1.

Table 1 Reedbed condition

Location	Area (ha)	Grid ref.	Condition	State	Main problem
1. Lytchett Bay	16.62	SY975925	Good	Stable except for…	Drying at edges
2. East Holton (Holton Lee)	35.26	SY965915	Good to poor	Declining	Sika Deer grazing
3. Holton Heath	17.31	SY950901	Very good	Stable	Minor grazing, some scrub
4. Keysworth reedbeds (five sub-areas)	Total: 25.5	SY934886 east to SY947896	Good	Stable	Could improve water control in main bed
5. Swineham Point	9.01	SY944878	Very good	Stable	Erosion at harbour edge
6. The Moors	22.53	SY953874	Good	Stable	Minor deer damage
7. Slepe	12.77	SY960865	Good	Stable to slight decline	Saline influx
8. Salterns	10.77	SY966869	Good	Stable	Saline influx, minor deer damage
9. Middlebere	7.1	SY965868	Good	Part declining	Cattle damage
10. Middlebere - Wych Lake west	.61	SY967853	Poor	Declining	Drying out
11. South Middlebere	9.87	SY961847	Good	Slow decline	Scrub invasion
12. Wych Lake east	1.57	SY986857	Good to poor	Declining in parts	Cattle grazing
13. Brownsea Island	5.63	SY028878	Very good to average	Reedbed 2 declining	Slow drying out in 2, minor deer damage elsewhere
Total area	174.55				

Locations 2, 9, 10, 11, 12 and part of 13 require immediate management. Others need a management plan and/or work, though no urgency is required.

The Poole Harbour reedbed resource is impressive and of great value to wildlife including many notable species. Generally the structure of the Poole Harbour reedbeds was good, the condition of the Holton Lee bed being the only one of serious concern because of the deer damage. Brownsea 1, Salterns, Slepe and South Middlebere had the greatest diversity with flowing freshwater, associated mires and carr and good reed condition. Holton Heath and Lytchett Bay possess the largest expanses of dense extensive reed cover. In the mainly tidal/saline areas, reeds grow tall, strong and dense, often becoming weaker and curved towards dry land.

Hydrology

Most of the Poole Harbour reedbeds are tidal with varying levels of salinity. Many have some freshwater areas and a few are purely freshwater. Saline beds are important for several key species including Bearded Tits, but will be less diverse than a similar freshwater bed, which will have a wider variety of invertebrate interest and thus more potential for feeding birds, Water Voles and a mix of marsh flora.

Most harbour tidal reedbeds do not need any significant management to maintain their existing state (excluding deer and grazing problems). Tidal beds generally had less leaf litter where they were regularly inundated. Freshwater beds have the deepest leaf litter and scrub encroachment.

Purely freshwater reedbeds are at South Middlebere, Middlebere - Wych Lake west, and Brownsea Island reedbed 1. Most other beds have a freshwater influence from the nearby higher ground in the form of flushes, mires and flowing water. Middlebere, Salterns, Slepe and Brownsea Islands 2 and 3 reedbeds all have a significant percentage of freshwater areas.

Poole Harbour reedbeds and wildlife

The particular benefit of reedbeds to wildlife results from their physical structure, i.e. large areas of dense reeds growing to around 2 m in height with about 100–200 stems per square metre and subject to regular flooding, difficult to penetrate and with a limited and specialized food source seasonal in availability. Each year new stems grow and the previous year's stems persist as tough vertical straws giving the bed a characteristic golden colour until about midsummer, when the new green growth of that year's reeds pushes upwards. Pools, ditches, scrub and rushy areas or saltmarsh may add variety but reedbeds are generally quite monotonous habitats.

Only a few species are specialized enough to cope with these conditions, for example:

- Reed Warblers and Bearded Tits use the stems as nest supports
- Water Rails nest in tussocks and feed on invertebrates in the reedbed
- Water Voles eat young reed shoots as their main diet for much of the year

- a few insects, such as Wainscot Moths, are so specialized that their whole life cycle occurs in just a few square metres of a reedbed, even wintering in the hollow reed stems

several of these specialist species are now rare or declining.

Dorset's threatened reedbed wildlife

Once more extensive reedbeds and adjacent marshes across the country supported a rich and interesting wildlife. Now they are relatively small, often invaded by scrub and separated by vast expanses of farmland causing some of these wildlife populations to diminish. Examples include the Bittern that demands large freshwater reedbeds and the Water Vole that survives best within a continuous network of wetland habitats. Other species, such as Wainscot Moths, are secure in very local areas. Future losses to their habitat by mismanagement or rising sea levels could bring the Dorset populations of reedbed species to local extinction.

Tables 2–6 show some of the key species that have been recorded in the Poole Harbour reedbeds over the last decade. Many are nationally or locally scarce and are associated almost solely with the reedbed habitat.

Status of Purbeck reedbed species

Of the six nationally rare Red Data bird species cited in the *UK Habitat Action Plan for Reedbeds*, Marsh Harrier, Cetti's Warbler and Bearded Tit are regular users of the Purbeck reedbeds. The Bittern (only rarely recorded in Poole Harbour) and Marsh Harrier (once breeding and now usually only seen in the winter) are on the RSPB's 'Red' list (high conservation concern). The Bearded Tit and Cetti's Warbler (found breeding in only a few other sites around the UK) are on the 'Amber' list (of medium conservation concern). The RSPB surveyed Cetti's Warbler in Dorset in 1996 with the following results (Table 5).

Table 2 Invertebrates more or less limited to reedbeds or associated habitats

A spider	*Clubiona juvenis*
A stonefly	*Nemoura dubitans*
Flame Wainscot Moth	*Senta flammea*
Brown-veined Wainscot	*Archanara dissoluta*
Twin-spotted Wainscot	*Archanara gemipunctata*
Silky Wainscot	*Chilodes maritimus*
Obscure Wainscot	*Mythimna obsoleta*
Small Red Damselfly	*Ceriagrion tenellum*
Downy Emerald Dragonfly	*Cordulia aenea*
Ruddy Darter Dragonfly	*Sympetrum sanguineum*

Table 3 Bird species more or less limited to reedbeds or associated habitats

Marsh Harrier	*Circus aeruginosus*
Reed Warbler	*Acrocephalus scirpaceus*
Cetti's Warbler	*Cettia cetti*
Sedge Warbler	*Acrocephalus schoenobaenus*
Aquatic Warbler	*Acrocephalus paludicola*
Reed Bunting	*Emberiza schoeniclus*
Bearded Tit	*Panurus biarmicus*
Eurasian Bittern	*Botaurus stellaris*

Table 4 Some species, common and scarce, often found in reedbeds but not exclusively

Hen Harrier	*Circus cyaneus*
Water Rail	*Rallus aquaticus*
Moorhen	*Gallinula chloropus*
Common Snipe	*Gallinago gallinago*
Jack Snipe	*Lymnocryptes minimus*
Stonechat	*Saxicola torquata*
Whinchat	*Saxicola rubetra*
Water Vole	*Arvicola terrestris*

Table 5 Cetti's Warbler survey 1996

Location	Numbers of singing males
Arne Moors/Ridge Farm (the Moors)	5–6
Keysworth	6–7
Lytchett Bay	2
Sherford River	2–3
Middlebere/Wych Channels	1
Swineham/Bestwall	12–13

Source: *British Birds*, March 1998.

Old records exist for the Reed Leopard Moth (a Red Data book species) in the Purbecks. In recent years other scarce moths, the Flame-, Obscure- and Blair's Wainscots have been found in some of the reedbeds. The Water Vole, though widespread, has suffered dramatic reductions in populations over the last 20 years, yet still survives in many of the freshwater reedbeds around the Purbecks.

The value of the reedbeds as roosting sites has been demonstrated by local bird watchers at Lytchett Bay (Table 6).

Table 6 Sample high counts

Passage migrant	Year	Reedbed roost during	Peak
Sand Martin	1995	July	2200 on 13th
	1996	July	c. 5000 on 18th
	1997	July	2300 on 14th
Swallow	1995	August	350 on 16th
	1999	Autumn	540 on 25th September
Pied Wagtail	1996	October	200 on 20th
	1999	October	200 on several dates

Species lists and management plans

Few of the reedbeds had adequate recording of the wildlife or detailed management plans. Lytchett Bay, Holton Heath, Keysworth and Brownsea Island have some good species records. Others have some but these tend to be included with records for the whole area and include records for non-reedbed species. Much work still needs to be done to provide up-to-date records of reedbed wildlife.

A detailed management plan exists for the Brownsea beds. Holton Heath, East Holton, the Moors, Salterns and Slepe all have some management statements or plans though little work was scheduled at the time of the survey. Others have no management plans or policies. However, East Holton has put in new plans to reduce problems and Slepe, Salterns and the Moors were having management reviews at the time of the survey.

Recommendations

General

- The area of reedbed in Poole Harbour should be maintained at the current area and future work should seek to increase it (within the Biodiversity Action Plan targets) for wildlife conservation reasons and for any losses associated with rising sea levels.
- Efforts should be made to secure the good management of reedbeds that are suffering from damaging activities.
- Grazing by deer and cattle should be examined and monitored. Beds with severe grazing problems will need to be protected from the grazing.
- The Reedbed Working Group should further seek to improve the perception and status of the Poole Harbour reedbeds by informing land managers, landowners and interested parties about survey results and management opportunities.

Future management work

- Management plans for all reedbeds should be drawn up to include managing peripheral habitats such as carr woodland.
- All reedbeds with surrounding carr woodland (all excluding Keysworth B to E, Swineham, the Moors, Middlebere and Wych Lake east) would benefit from coppicing/pollarding, scrape and pool creation and other reedbed margin improvement works.
- Options for buffer zones and the creation of more scrub along the margins of all reedbeds would benefit biodiversity and specifically aid Cetti's Warblers.
- All efforts should be made to maintain freshwater beds or freshwater components of beds where this is sustainable. An assessment of the values and future of existing or new sluices, banks and ditches should be done and sluices brought back into action where this fits in with conservation management.
- Scrub encroachment should be addressed where it occurs.

Opportunities for expansion

Possible opportunities for reedbed expansion which may yield significant amounts of new wetland/reedbeds are at: the Moors (work already in hand by the RSPB), Keysworth and South Middlebere. Other opportunities may exist at Lytchett Bay south, Swineham and Holton Lee (north) and the lower Rivers Frome and Piddle. These are only theoretical and depend upon land ownership, land values to stock and other conservation issues.

Surveying and recording

- All beds should be surveyed for breeding and wintering birds and a longer term approach to monitoring developed.
- Records held at the Dorset Environmental Records Centre (DERC) need sorting and those applicable to reedbeds need identifying.
- Specialists must be encouraged to research in reedbeds, especially for invertebrates.
- More searching at, e.g. Centre for Ecology and Hydrology, and private individuals could yield more species information.
- There are other reedbeds in Poole Harbour and its reaches that could be surveyed.

Acknowledgements

Particular thanks must go to the following: The Reedbed Working Group who steered the project and whose parent organizations funded the survey: Paul St Pierre and Dr John Day (RSPB), John Stobart (English Nature), and Emma Rothero (Environment Agency). I would also like to thank Nick Squirrel (English Nature), Annabel King (National Trust) and Richard Acornley (Environment Agency).

All landowners who allowed access to their land including the Wimborne and Rempstone Estates, Christopher Lees and the Lee Estate, the Holton Lee Trust, Harry Clark of Keysworth Farm, Camus Aggregates, the RSPB, the National Trust, English Nature and the Dorset Wildlife Trust.

Records, area figures and maps were kindly supplied by DERC (particular thanks to Alison Stewart and Nick Button), bird records by Shaun Robson, Bob Gifford, Tasie Russell and Steve Smith and moth records by Alan Bromby and Peter Davey.

References

Cook. K. (2001) Reedbeds of Poole Harbour. *Poole Harbour Study Group Report*, No.1. Wareham: Poole Harbour Study Group.

6. History and Ecology of *Spartina anglica* in Poole Harbour*

Alan Raybould

Syngenta, Jealott's Hill International Research Centre, Bracknell, Berkshire RG42 6EY
e-mail: alan.raybould@syngenta.com

Spartina anglica is a grass of saltmarshes and mudflats. It resulted from chromosome doubling in a hybrid between a native grass, *S. maritima*, and a species accidentally introduced from North America, *S. alterniflora*. The hybrid, *S. x townsendii*, was first recorded in 1870 near Hythe in Southampton Water; *S. anglica* was first recorded in 1892 from Lymington. *Spartina anglica* arrived in Poole Harbour in the 1890s, and by the early 1920s it covered over 800 ha of mudflats that had previously been largely clear of vegetation. During the late 1920s, *S. anglica* began to recede, and by 1980 about 350 ha remained. The rapid spread and decline of *S. anglica* was associated with physical and biological change within the harbour, the most noticeable being changes in the depth of navigation channels and the colonization of pure swards of *S. anglica* by other saltmarsh plants. The spread of *S. anglica* is also implicated in local reductions in the populations of wading birds.

Introduction

Poole Harbour is one of the most important features in the natural history of Dorset. It was formed when a post-glacial rise in sea level submerged land surrounding a 'Solent River', which flowed eastwards across land that is now Poole Bay (Bird and Ranwell, 1964). The Rivers Frome and Piddle, which flow into Poole Harbour at Swineham, are the headstreams of that river. The area of Poole Harbour was greatest about 6000 years ago, since when natural sedimentation and land reclamation have reduced the area by about 1000 ha to the present 3600 ha (May, 1969; Gray, 1986a).

Although urban development has affected the edges of the harbour, a probably more significant change occurred within the body of the harbour in the last 100 years by natural means. In the 1890s, mudflats began to be colonized by a perennial grass called *Spartina anglica* (hereafter *Spartina* unless stated otherwise). *Spartina* spread very quickly in the harbour and covered over 800 ha by 1924. For various reasons, there has since been much loss of *Spartina*, and the species now covers less than 400 ha (Gray *et al.*, 1991). This chapter examines some of the effects that the changes in the distribution of *Spartina* have had on the hydrography and ecology of Poole Harbour.

*First published in the *Proceedings of the Dorset Natural History and Archaeological Society.*

The origin of *Spartina anglica*

Spartina anglica is a textbook example of the evolution of a new species by allopolyploidy, the hybridization of two species to give a fertile new species with about twice the mean chromosome number of the parental species. An allopolyploid can be formed either by the fusion of diploid gametes or by chromosome doubling in a (usually sterile) F_1 hybrid.

The '*Spartina* story' (Lambert, 1964) began with the accidental introduction of *Spartina alterniflora*, a plant of the eastern seaboard of North America, into Southampton Water sometime before 1816. At Hythe, and at the mouth of the Itchen, *S. alterniflora* grew close to *S. maritima*, a native plant that is now extinct in Southampton Water (Townsend, 1883; Raybould *et al.*, 1991b). In about 1870, a new form of *Spartina* was collected at Hythe. This *Spartina* spread slowly in Southampton Water until the late 1880s when "something occurred that favoured the spreading of the grass" (Stapf, 1913). All the morphological, cytological and biochemical evidence (Marchant, 1967, 1968; Raybould *et al.*, 1991a) suggests that the new *Spartina* was a sterile hybrid between *S. alterniflora* and *S. maritima* (with *S. alterniflora* as the female parent (Ferris *et al.*, 1997)) and the 'something that favoured the spreading' was chromosome doubling in the hybrid, which produced a fertile allopolyploid able to spread by seed.

The F_1 hybrid is named *Spartina* x *townsendii*, and is still abundant at Hythe and also occurs in small patches scattered throughout the British Isles. The fertile allopolyploid is named *S. anglica*. This species spread along the south coast of England and the north coast of France by the natural dispersal of seeds and plant fragments (Oliver, 1920, 1925). *Spartina anglica* was planted extensively on mudflats in other parts of the British Isles, in north-west Europe and in several other countries (notably Australia and China) for coastal defence and land reclamation (Ranwell, 1967).

The spread of *Spartina anglica* in Poole Harbour

Hubbard (1965) gives 1890 as the date of the first appearance of *Spartina* in the harbour, although this is an estimate based on anecdotal evidence in the personal correspondence of local residents. Oliver (1925) gives 1899 as the first confirmed record. Seedlings and plant fragments of *Spartina* established on driftlines or flat mud by being trapped in mats of eelgrass (*Zostera* spp.) (Hubbard, 1965). Once established, plants spread vegetatively at "several feet" per year (Oliver, 1925), particularly in the bays and inlets of the southern shore west of Ower (Hubbard, 1965). The leeward sides of islands and mudflats in the Wareham Channel were also colonized rapidly. The most dramatic spread was in Holes Bay, where 63% of the intertidal zone became covered with *Spartina* between 1901 and 1924 (Gray and Pearson, 1984). On the north shore of the harbour to the east of Holes Bay, there was little growth of *Spartina*, apart from sheltered sites in Parkstone Bay (Hubbard, 1965). In general, towards the mouth of the harbour, *Spartina* remained as disconnected clumps and in the centre of the harbour, swards formed on the fringes of bays. Only in the upper reaches of the harbour, west of a line from Fitzworth Point to Hamworthy, was there extensive sward formation (Hubbard, 1965).

The seaward limit for successful establishment of *Spartina* in Poole Harbour was about Ordnance Datum (Newlyn) (Hubbard, 1965). This limit is much lower than other estuaries on the south and west coasts of Great Britain and is probably due to the small spring tidal range in the harbour (Gray *et al.*, 1991). The upper limit varies between +0.5 m OD (e.g. in Brands Bay) and +0.9 m OD (e.g. Keysworth and Arne Bay) (Ranwell *et al.*, 1964).

Spartina reached its maximum area in Poole Harbour between 1917 and 1924, at which time it covered about 800 ha (Gray *et al.*, 1991). However, by 1919, some *Spartina* marsh was being eroded, possibly due to the migration of a creek (Oliver, 1920). The rate of loss increased during the 1920s as marshes in exposed locations were subject to tidal scour (Oliver, 1925). Marshes either side of Wych Channel at Shipstal Point were particularly affected in the early 1920s (Hubbard, 1965).

Between 1924 and 1952, there was some spread of *Spartina* on the south-west shore of the harbour and along the shores of the Wareham Channel. However, the general trend was recession of *Spartina*, particularly along the south-east shore and areas around Brownsea Island and in Holes Bay (Table 1). Hubbard (1965) estimated that there was a net loss of 172 ha of *Spartina* marshes between 1924 and 1952. Gray *et al.* (1991) used different methods and estimated 200 ha of *Spartina* were lost from the harbour between 1924 and 1952.

Gray and Pearson (1984) estimated that 250 ha of *Spartina* were lost between 1952 and 1980, leaving about 350 ha. The most severe recession occurred in areas east of the Arne peninsula, particularly around Furzey Island, to the west of Green Island and in Brands Bay. Smaller reductions took place around Long and Round Island and in the Middlebere Channel. *Spartina* was also lost from Holes Bay during this period (Table 2).

Recent figures are available for Holes Bay only (Table 2) and show *Spartina* is still receding. The main changes between 1981 and 1994 are the reduction in area of the intertidal area because of reclamation for the Holes Bay road, and continued fragmentation of the *Spartina* marsh on the spit extending south into the bay from the centre of the railway causeway.

There are several reasons for the loss of *Spartina* in Poole Harbour. The first is erosion at the edges of marshes. This began in the 1920s (see above) and is still continuing, being particularly prevalent between Holton Heath and Rockley Sands. Gray and Pearson (1984) found that loss of *Spartina* between 1952 and 1980 was due to the break-up of marsh on intertidal mudflats, rather than on the marshes fringing land.

Another important cause is 'die-back', where *Spartina* degenerates in patches in the body of a sward, rather than at an eroding edge. The cause of die-back is still not known completely, although the process seems to be associated with badly drained, highly

Table 1 Estimated loss/gain in area of *Spartina* in Poole Harbour

	Date of establishment (E) and/or sward completion (S)	1924	1952	Loss/gain
Brands Bay	S 1914	75.5	80.4	+4.9
Whitley Lake	E 1901	1.2	4.0	+2.8
Furzey and Green Island		76.4	45.0	-31.4
Southern Shore (east)	S 1914	101.0	73.4	-27.6
Southern Shore (west)	E 1890 S 1913–14	102.1	123.2	+21.1
Parkstone Bay	E 1901	37.1	2.2	-34.9
Adjacent to Brownsea Island		52.3	0	-52.3
Long and Round Island and Grip Heath	S 1914	67.8	53.8	-14.0
Arne Bay	E 1898 S 1915	30.2	40.6	+10.4
Holton - Rockley	E 1907	24.7	26.8	+2.1
Holton Mere (Keysworth)	E 1912	26.4	43.8	+17.4
Giggers' Island and Swineham	E 1915	0.4	11.9	+11.5
Holes Bay	E 1901 S 1924	227.7	140.4	-87.3
Lytchett Bay	E 1910	44.1	43.8	-0.3
Brownsea Island	E 1930	0	5.4	+5.4
Total		866.9	694.7	-172.2

Source: Hubbard (1965).

Table 2 Area of Poole Harbour covered by *Spartina*

Date	Area in whole harbour	Area in Holes Bay (proportion of intertidal)
1924	800	208 (0.63)
1952	600	132 (0.41)
1972		95 (0.32)
1980	350	80 (0.29)
1994		63 (0.25)

Source: Pre-1990 data from Gray *et al.* (1991) and Gray and Pearson (1984) who used slightly different methods from Hubbard (1965).

anaerobic soils in which the *Spartina* rhizomes may be poisoned by sulphide ions and lack of oxygen (Goodman and Williams, 1961; Gray *et al.*, 1991). Die-back also began in Poole Harbour in the 1920s and is thought to be the principal cause of the large reduction of *Spartina* in Holes Bay (Hubbard, 1965). Die-back is still occurring in parts of the harbour, in Brands Bay, for example.

Finally, *Spartina* marsh has been lost through invasion by other species from the landward edge. Invasion is characteristic of marshes to the west of Arne (Gray and Pearson, 1984), where large areas have been replaced by *Phragmites communis* (common reed) in areas with low salinity (Ranwell *et al.*, 1964; Gray, 1986a). Other species that have invaded the *Spartina* sward at Keysworth are *Scirpus maritimus* (Sea Club-rush), *Elytrigia atherica* (Sea Couch Grass), *Agrostis stolonifera* (Creeping Bent), *Festuca rubra* (Red Fescue) and *Puccinellia maritima* (common saltmarsh grass) (Hubbard and Stebbings, 1968). *Spartina* swards have also been invaded by *Atriplex portulacoides* (Sea Purslane) and *P. maritima* in, for example, Parkstone Bay, Lytchett Bay, Brands Bay and in Middlebere and Wych Channels (Gray, 1986a).

Poole Harbour is the source of much of the *Spartina* planted in the British Isles and abroad for coastal defence and land reclamation. In the 1920s, 40,000 plants of *Spartina* were exported from Poole to Holland for reclamation following a successful trial of 50 plants by the Dutch government. In 1929, the Ministry of Agriculture published a pamphlet on the uses of *Spartina* which generated world-wide interest and many requests for plant material. Up to 1936, over 35,000 plants were exported from Manningtree (Essex) populations that originated from plants taken from Poole Harbour in 1924. Between 1928 and 1936, over 85,000 plants were exported directly from Poole to sites in Ireland (16,000 plants to 12 sites), the UK (30,000 plants to 26 sites), Germany (20,000 plants), Denmark (7000 plants) and several other countries including Australia and Trinidad. Hubbard (1965) estimated that between 1924 and 1936, over 175,000 plants and many seed samples were sent from Poole to at least 130 sites around the world. The outcome of these introductions is described by Ranwell (1967).

Most of the *Spartina* in Poole Harbour is the fertile allopolyploid, *S. anglica*. However, the sterile F_1 hybrid, *S.* x *townsendii* has also been found. In the 1960s, Hubbard (1965) recorded *S.* x *townsendii* from the landward side of marshes in Arne Bay, to the east of Fitzworth Point, at Ower, from both sides of Goathorn Point and Keysworth. Small patches were also found at Holes Bay, Brands Bay, Whitley Lake, Furzey Island and Green Island. Isolated tussocks also occurred on the seaward side of the Keysworth marsh. No systematic search has since been made for *S.* x *townsendii*, although clones have been found in Brands Bay with very high tiller density, which is characteristic of this species. The *S.* x *townsendii* plants in Holes Bay have almost certainly been lost following reclamation for development at Fleetsbridge and for the Holes Bay road. It is notable that *S.* x *townsendii* was particularly prevalent at Arne, because it is from this area that much *Spartina* was exported. The dispersal of *Spartina* from Poole may explain the occurrence of *S.* x *townsendii* in Norfolk and Ireland (Hubbard, 1965).

Hydrographical change associated with *Spartina*

Spartina has greatly affected the sedimentation regime of the harbour (for detailed accounts see Bird and Ranwell, 1964; Ranwell, 1964; Hubbard and Stebbings, 1968), first by accreting and consolidating sediment by rhizome growth, and then by releasing this sediment as the sward eroded and died-back. In its period of expansion, *Spartina* accreted sediment rapidly. Hubbard and Stebbings (1968) estimated that locally in the upper harbour up to 1.8 m of sediment was trapped by *Spartina*. In other parts of the harbour, depths of 70–100 cm were common, dropping to 35 cm near the harbour mouth.

Only a small proportion of the sediment accreted by *Spartina* entered the harbour during the accretion process. Hubbard and Stebbings (1968) estimated that about 1 million tons of sediment were trapped by *Spartina* in the upper harbour over 55 years. During this time, the amount of sediment brought into the harbour by the River Frome was about 40,000 tons (1/25th of the total). The main hydrographical effect of the expansion of *Spartina* was, therefore, the stabilization of previously mobile sediment, rather than the accretion of new material. This stabilization led to a deepening of many navigation channels between the 1849 and 1934 hydrographical surveys of the harbour (Figure 1 and Table 3).

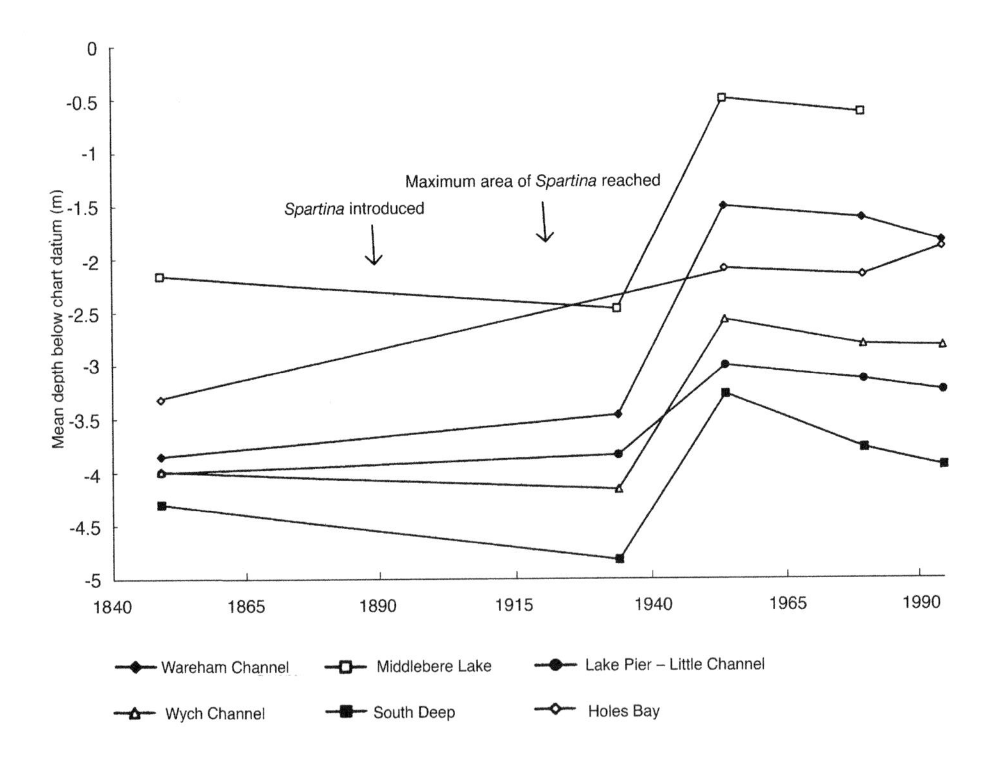

Figure 1 Changes in the mean depth of channels in Poole Harbour.

Table 3 Depths of navigation channels in Poole Harbour

	Mean channel depth (m)				
	1849	1934	1954	1980	1995
Main harbour					
Lake Pier - Little Channel	4.00	3.85	3.02	3.16	3.28
Middlebere Lake	2.16	2.48	0.51	0.65	*
South Deep	4.30	4.83	3.29	3.80	3.98
Wareham Channel	3.85	3.48	1.52	1.64	1.87
Wych Channel	3.99	4.17	2.59	2.83	2.86
	1849		1952	1980	1995
Holes Bay - Backwater Channel					
East Arm	2.49		1.09	1.67	0.72
West Arm	3.55		1.26	0.99	0.76
South Arm	3.50		3.43	3.46	3.88

*No survey as channel too shallow.

Source: Pre-1995 *Spartina* cover from Gray *et al.* (1991).

The die-back and erosion of *Spartina* caused an enormous amount of sediment to be released back into the harbour quickly (Green 1940). Between 1934 and 1954, there was considerable shoaling in all channels in the main harbour (Figure 1). In the long channels of the north and west part (Wareham and Wych), shoaling was greatest in the upper parts. For example, the average depth change in the lower 7000 feet of Wareham Channel was 3.17 m, while in the upper 4000 feet the average change was only 1.48 m (Figures 2 and 3). There was no survey of Holes Bay in 1934 but we can infer from the changes in other parts of the harbour that shoaling between 1849 and 1954 (Table 3) may have begun in the 1920s as the *Spartina* regressed.

Since 1954, the channels have deepened in most parts of the harbour (Figure 1 and Table 3), although the dredging of some of these channels was considered in the early 1980s (Poole Harbour Commissioners, 1982). South Deep may have deepened the most due to greater scouring near the harbour entrance. This effect is seen in miniature in Holes Bay where the south arm of Backwater Channel at the mouth of the bay deepened between 1954 and 1995, while the east and west arms shoaled (the deepening in the east arm between 1954 and 1980 was probably the result of land reclamation for road construction) (Figure 4).

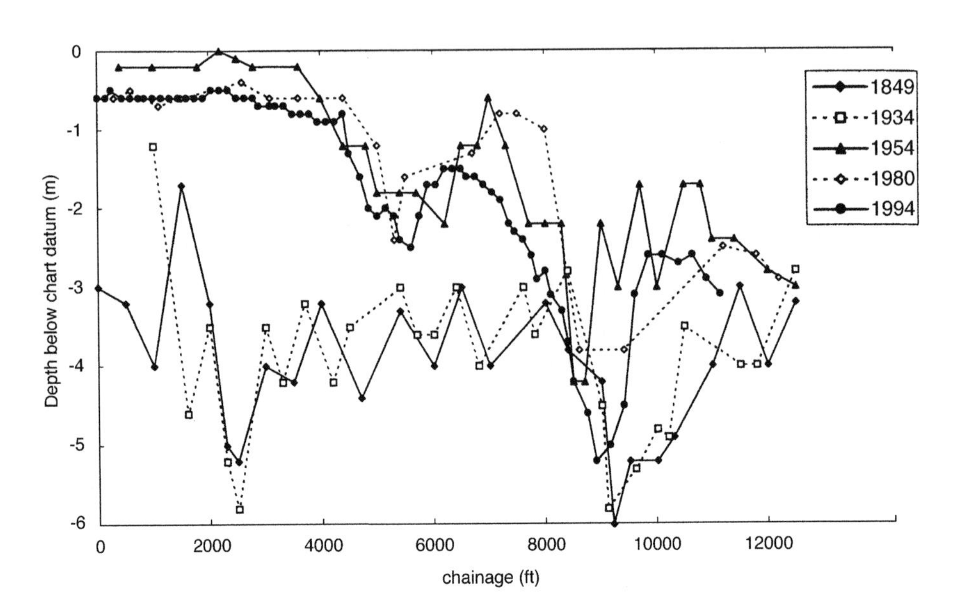

Figure 2 **Changes in the profile of Wareham Channel. The chainage line runs from just north of Gigger's Island (0 feet) to the shore just north of Lake Pier (12,000 feet).**

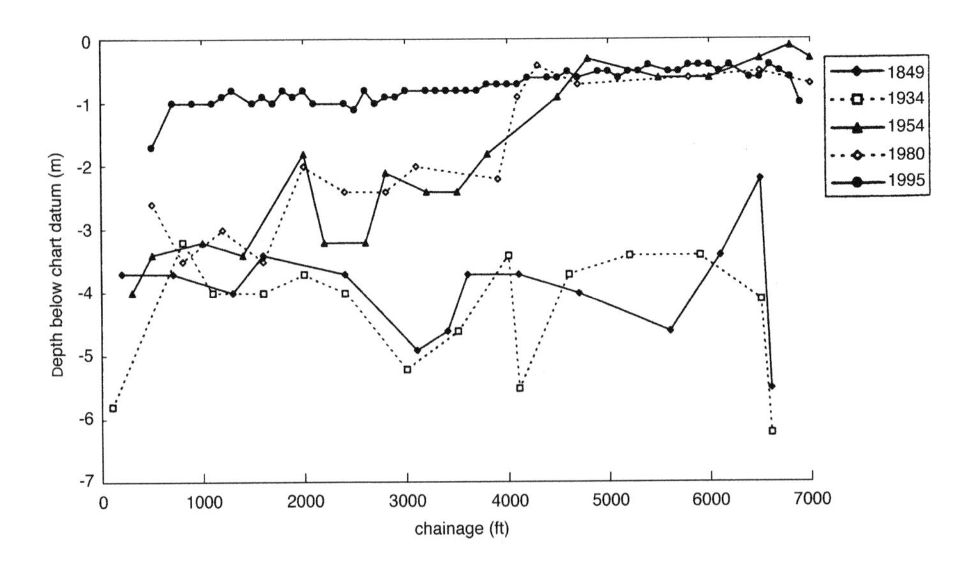

Figure 3 **Changes in the profile of Wych Channel. The chainage line runs from east of Brownsea Island (0 feet) in a north-easterly direction and from 4000 feet runs east to Shipstal Point (7000 feet).**

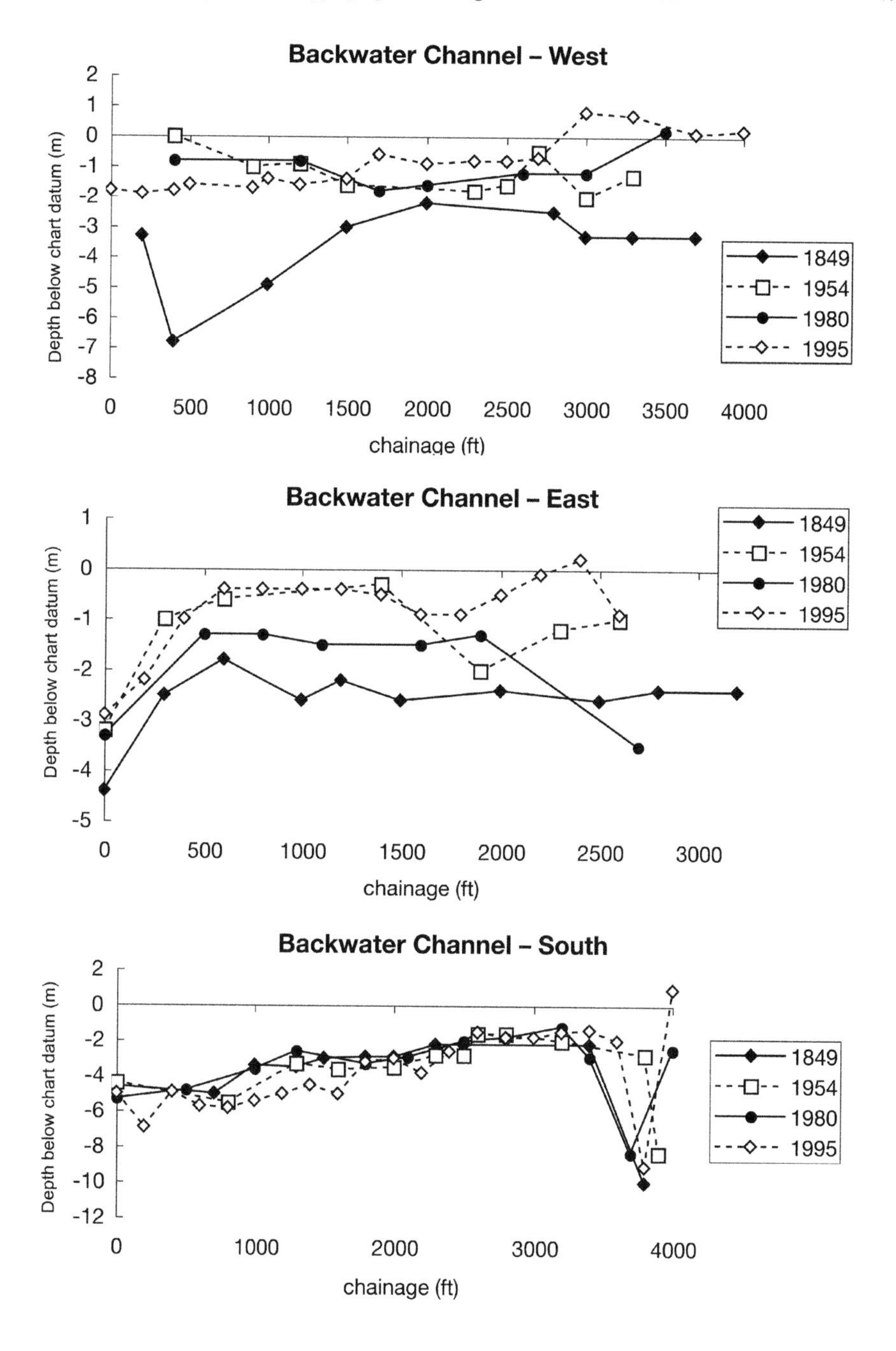

Figure 4 **Changes in the profiles of channels in Holes Bay. The south chainage line runs from Poole Bridge (0 feet) to the centre of the bay. The east and west chainage lines are 'L' shaped running from the mouth of the south channel to the railway causeway.**

An exception to the general deepening of the harbour is Upper Wych Channel (Figure 3). Between 1980 and 1995, the average depth of the channel changed from 1.7 m to 0.75 m. The effect was most marked in the lowermost 4000 feet (chainage 0 to 4000 feet) where the average depth was 2.6 m in 1980, but only 0.9 m in 1995.

In general, estuaries such as Poole Harbour reduce in volume over time as they act as 'settlement tanks'. The long-term trend in the channel depths before the introduction of *Spartina* was, therefore, probably a slow shoaling. As the sediment from eroding *Spartina* marshes is redistributed, the deepening may reverse and shoaling may begin again. In the future, the perturbation of channel depths caused by *Spartina* may be viewed as a blip in a long-term trend. Nevertheless, *Spartina* has had a remarkable influence on processes normally determined by large-scale physical factors.

Ecological change associated with *Spartina* in Poole Harbour

Vegetation changes

Throughout most of its range, *S. anglica* is found in a zone immediately seaward of other saltmarsh communities (Gray *et al.*, 1991). *Spartina* was able to colonize this 'vacant niche' (Gray, 1986b) because of its greater ability to oxidize toxins (e.g. sulphides and ferrous compounds) and to tolerate salinity, flooding and sediment accretion (Gray *et al.*, 1991). The pattern of spread in Poole Harbour (Hubbard, 1965) is consistent with colonization of bare mud, followed by invasion by other species from the landward edge of the sward.

Before the arrival of *S. anglica*, *Puccinellia maritima* was the commonest perennial plant in the lowest zones of saltmarshes in western Europe. In some areas, *Spartina* invaded *Puccinellia*, while in others *Puccinellia* invaded *Spartina*. In Great Britain, *Puccinellia* generally replaces *Spartina* in north-western saltmarshes (e.g. in Morecambe Bay). This effect is at least partly a result of the species' different mechanisms of photosynthesis, which cause *Puccinellia* to begin spring growth earlier at low temperatures and shade emerging *Spartina* shoots (Scholten and Rozema, 1990). The change from clayey and muddy substrate in the south and east to sandy substrates in the north-west is also a factor in the latitudinal change in the interaction between *Spartina* and *Puccinellia* (Scholten and Rozema, 1990). *Puccinellia* has invaded some *Spartina* marshes in Poole Harbour (e.g. Keysworth and east Lytchett Bay, Gray, 1986a), although the pattern of *Puccinellia* invasion is probably not associated with substrate variation (A. J. Gray, pers. comm.). The commonest other plant growing within *Spartina* is *Atriplex portulacoides,* which has invaded marshes in Parkstone Bay, Lytchett Bay, Brands Bay and Middlebere (Gray, 1986a; and see above).

At Keysworth, *Spartina* did more than just colonize bare mud. Studies of the stratigraphy of the area by Hubbard and Stebbings (1968) showed that the area was colonized successively by *Potamogeton pectinatus* (Fennel Pondweed), *Ruppia cirrhosa* (Spiral Tasselweed), *R.*

maritima (Beaked Tasselweed) and *Zostera* spp. (eelgrasses). The succession of these species suggests a gradual transition from freshwater to sea water, associated with a rise in sea level and/or a drop in land level. *Spartina* was found at -0.96 m OD beneath the oldest part of the Keysworth marsh, which is about 1 m below the present lower limit. This suggests that there was considerable compaction of the marsh substrate, causing the remains of the initial colonization to sink to a much lower level (Hubbard, 1965).

The effect of *Spartina* on wading birds

Poole Harbour is the most important estuary for wildfowl in Dorset, and supports nationally important populations of several species. The Black-tailed Godwit and Shelduck populations are of international importance (Prater, 1981). Ducks generally feed in Brands Bay, Newton Bay and at Arne, although the Bar-tailed Godwit prefers sandier areas such as Sandbanks (Gray, 1986a). Waders feed in these areas and also in the Wareham Channel and Holes and Lytchett Bays. As with other estuaries in southern England, double high tides in Poole Harbour tend to reduce the feeding time available to intertidal birds, compared with similarly sized areas in other parts of the country (Prater, 1981).

The spread of *Spartina anglica* on mudflats has been implicated in the decline of wader (particularly Dunlin *Calidris alpina*) populations in many British estuaries (Goss-Custard and Moser, 1988; Davis and Moss, 1984; Millard and Evans, 1984) because their invertebrate prey species decline and/or are less accessible within *Spartina* swards compared with bare mudflats. However, Goss-Custard and Moser (1988) found no evidence of an increase in Dunlin numbers in areas where *Spartina* was receding. Indeed in the Solent, Dunlin numbers declined during the 1970s and 1980s in spite of *Spartina* decrease (Tubbs *et al.*, 1992).

One possible reason why waders have not increased in areas that have lost *Spartina* is a lag between *Spartina* recession and recolonization by wader prey species. Poole Harbour was one of the first estuaries to lose large areas of *Spartina* and so may be expected to be one of the first to show recovery of bird numbers if former *Spartina* marshes can revert to wader feeding grounds.

Peak wader counts for Poole Harbour for winters between 1969–70 and 1993–94 were obtained from the British Trust for Ornithology's annual reports of the Birds of Estuaries Enquiry (BoEE). The counts are shown in Figure 5. A rank correlation shows a highly significant upward trend in wader numbers ($r_S = 0.633$, $P < 0.01$). An exponential regression of number on year was also highly significant ($b = 0.0301$, $R^2 = 41.5\%$, $P < 0.001$) and showed an average annual percentage increase of 3.01% in wader numbers.

To test further the association between total wader numbers and *Spartina*, counts were obtained from BoEE reports for 14 other British estuaries which had continuous runs of data between 1971–72 and 1993–94. For each estuary the following calculations were made (Table 4):

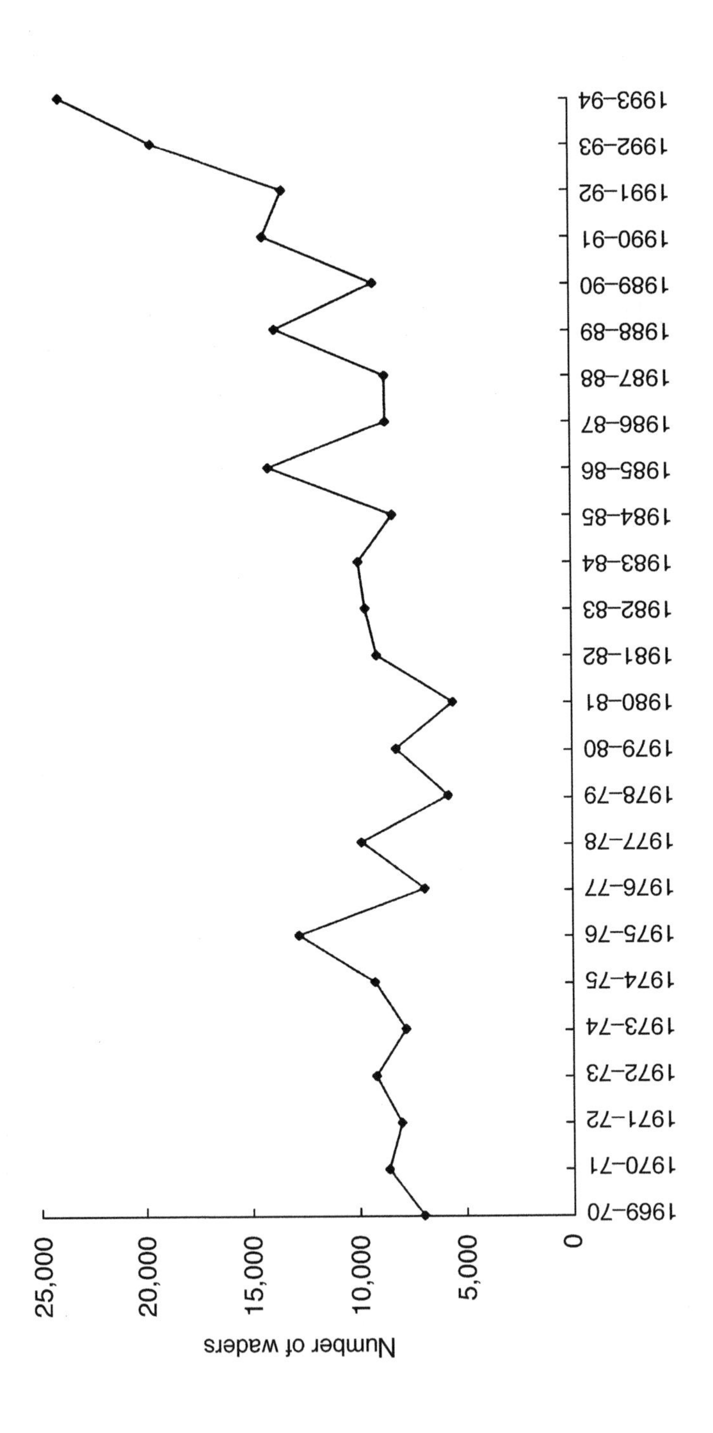

Figure 5 **Changes in the total number of waders wintering in Poole Harbour.**

Table 4 Changes in wader numbers and *Spartina* cover in Poole Harbour and other estuaries

Estuary	r_s	ARI %	Minimum count (thousands) (year)	Maximum count (thousands) (year)	*Spartina* area (1) (ha)	*Spartina* area (2) (ha)
Increased *Spartina* cover						
Exe	-0.388	-1.33	11 (1982–83)	26 (1969–70)	6	24
Chichester Harbour	-0.097	0.88	17 (1970–71)	45 (1989–90)	715	815
Severn	-0.522	-2.52	34 (1986–87)	131 (1973–74)	303	383
Morecambe Bay	-0.437	-1.36	105 (1982–83)	269 (1973–74)	185	368
Lindisfarne	-0.617	-3.19	19 (1984–85)	62 (1978–79)	36	127
Reduced *Spartina* cover						
Langstone Harbour	0.441	1.62	16 (1970–71)	46 (1989–90)	193	65
Medway	0.768	5.81	5 (1979–80)	54 (1992–93)	278	76
Blackwater	0.524	3.65	11 (1972–73)	78 (1992–93)	469	35
Stour (Essex)	0.718	3.09	14 (1971–72)	46 (1992–93)	149	119
The Wash	0.656	3.43	85 (1975–76)	353 (1992–93)	1914	138
Burry Inlet	-0.137	-0.76	18 (1978–79)	40 (1973–74)	405	335
Dee (Cheshire)	-0.267	-0.01	17 (1978–79)	158 (1992–93)	405	321
Humber	0.760	4.33	38 (1972–73)	127 (1989–90)	162	120
Ribble	0.060	-0.08	54 (1983–84)	219 (1973–74)	405	330
Poole Harbour	**0.633**	**3.01**	**6 (1980–81)**	**24 (1993–94)**	**600**	**350***

Spartina cover (1) from Hubbard and Stebbings (1967) and (2) from Burd (1989) except *from Gray *et al.* (1991).
r_s = rank correlation between winter peak wader numbers and time.
ARI = annual rate of increase of the winter peak wader count.

(i) rank correlation between bird number and year (tests for strength of trend)

(ii) exponential regression of bird number and year to find annual percentage rate of increase (tests for magnitude of trend)

(iii) the change in area of *Spartina* between the estimates of Hubbard and Stebbings (1967) and Burd (1989).

Estuaries were split into two groups, one for sites where *Spartina* had increased between the surveys of Hubbard and Stebbings and Burd (+*Sp*) and the other for sites where it decreased (-*Sp*). The differences in bird number trends between the categories were tested with a Mann-Whitney test, which ranks values and then tests whether the sum of the ranks in each category is significantly different from that expected if values had been assigned randomly to the categories.

There was a significant difference between the +*Sp* and -*Sp* categories in both the strength and magnitude of the positive trend in bird numbers. In the -*Sp* group were areas with the strongest positive trends (test of $+Sp = -Sp$ vs $+Sp < -Sp$ significant at $\alpha = 0.0042$) and the largest positive trends ($+Sp = -Sp$ vs $+Sp < -Sp$ significant at $\alpha = 0.042$). This relationship was not due to site latitude, because rank correlations of both coefficients and latitude were not significant ($P = -0.182$ and 0.122 for rank correlation and regression coefficients respectively; $N = 15$ in both cases). It is also worth noting that wader numbers declined in all estuaries with increased *Spartina* cover, whereas wader numbers increased nationally between 1987–88 and 1991–92 (Cayford and Waters, 1996). These results suggest an association between changes in *Spartina* cover and wader numbers, although changing *Spartina* cover is not necessarily the cause of the trends in wader numbers.

As suggested above, the lag between *Spartina* regression and recolonization of mud by wader prey species may be the reason previous studies found no increase in wader numbers in areas that had lost *Spartina.* McGrorty and Goss-Custard (1987) estimated the density of four wader prey species, *Hydrobia ulvae*, *Nereis diversicolor*, *Corophium* spp. and 'small bivalves' and calculated an invertebrate diversity index (IDI), a measure of the abundance and diversity of invertebrate prey species, for several sites in 1987.

The IDI for 10 sites is shown in Table 5. There is a statistically significant rank correlation between the IDI and the strength of the upward trend (significance of rank correlation coefficient) in wader numbers between 1971 and 1994 ($r_s = 0.728$, $N = 10$, $P = 0.05$–0.01) and the annual rate of increase in wader numbers over the same period ($r_s = 0.685$, $N = 10$, $P = 0.05$–0.01). McGrorty and Goss-Custard (1987) also found a significant relationship between the IDI and the annual rate of increase of Dunlin numbers between 1975 and 1986.

In conclusion, there was a significant increase in wader numbers in Poole Harbour over the 25 year period and there is circumstantial evidence that this was due to invertebrates recolonizing mudflats previously occupied by *Spartina*. The time between *Spartina* loss and recolonization by invertebrates may be over 20 years (McGrorty and Goss-Custard, 1987), which could explain why wader numbers do not increase immediately.

Table 5 Relationship between invertebrate prey species and change in wader numbers for 10 British estuaries

Estuary	r_s	ARI	IDI
Medway	0.768	5.81	79
Stour	0.718	3.09	75
Poole Harbour	**0.633**	**3.01**	**70**
Blackwater	0.524	3.65	70
Langstone Harbour	0.441	1.62	42
Chichester Harbour	0.097	0.88	75
Hamford Water*	0.067	0.70	63
Exe	-0.388	-1.33	63
South-west Solent*	-0.481	-1.31	58
Portsmouth Harbour*	-0.759	-4.89	46

*Data not continuous.

r_s = rank correlation between winter peak wader numbers and time.

ARI = annual rate of increase of the winter peak wader count.

IDI = invertebrate diversity index (see text).

Table 6 The frequency of ergot infection of *Spartina* inflorescences and spikelets in Poole Harbour

Year	Inflorescences infected (%)	Spikelets infected (%)
1983	36	–
1984	61	–
1985	80	16.1
1987	–	22.8
1988	85	–
1992	84	–
1995	71	23.6
1997*	90	23.0

Source: 1980s inflorescence data from Gray *et al.* (1990).

*Data from Brands Bay and Bramble Bush Bay only.

Spartina and the ergot fungus

Since the early 1980s, *Spartina* in Poole Harbour has become increasingly infected by the Ergot fungus *Claviceps purpurea* (Table 6). The life cycle has been described by Gray *et al.* (1990). The most obvious symptom of infection is the sclerotium or ergot, a mass of fungal hyphae which protrudes from an infected spikelet. The ergots overwinter in the surrounding mud and germinate in the spring, producing sexual ascospores which land on *Spartina* stigmas and lead to a primary infection. Secondary infection happens through dispersal of asexual conidiospores extruded from infected spikelets in a sticky honeydew. Development of sclerotia completes the life cycle.

The epidemic on *Spartina* in Poole Harbour is unusual, not so much for its high infection rate, but because of its persistence and uniformity. Other grass species such as *Lolium perenne* (Ryegrass) (Jenkinson, 1958) and *Spartina alterniflora* (Eleuterius and Meyers, 1974; Gessner, 1978) have high ergot infection rates, but these tend to be short-lived, confined to particular habitats within sites, or both. As Table 6 shows, the proportion of infected inflorescences in Poole Harbour remained at over 70% for 10 years. Also among 15 sites throughout the harbour (see Gray *et al.*, 1990), there is no tendency for differentiation into populations with consistently high or low infection rates (Friedman two-way analysis of ranks S = 16.28, d.f. = 11, $P = 0.133$) (Raybould *et al.*, 1998). There are no published long-term data sets for ergot infection of other *Spartina anglica* populations. However, Gray *et al.* (1990) reported observations of Thompson on the Dee estuary marshes, which suggest that a persistent infection was developing in the late 1980s. It seemed, however, that infection rates were consistently higher in the mature zone compared with the pioneer zone.

The probable reasons for the high, persistent and uniform infection rates in Poole Harbour are the genetic uniformity of *Spartina* (Raybould *et al.*, 1991a), the lack of zonation of *Spartina* in the harbour, the closed nature of the harbour allowing the build up of inoculum, and perhaps the old age of many clonal lines (c.f. the mature zone on the Dee). The reduction in infection rate in 1995 compared with 1992 may be due to the hot summer of 1995 as ergot spread is favoured by cool, damp conditions (MAFF, 1974).

It is difficult to estimate whether the ergot is having a significant deleterious effect on *Spartina*. The most obvious potential effect is reduction in seed production. Raybould *et al.* (1998) found that in 1985 and 1995, there was no significant difference between the average number of seeds produced on uninfected and infected inflorescences. Although inflorescences with high ergot infection (more than 10% of inflorescences infected) produced fewer seed than uninfected inflorescences, the lower seed output was offset by inflorescences with light infection (fewer than 10% of spikelets infected) producing more seed compared with uninfected inflorescences. As Raybould *et al.* (1998) point out, these results do not necessarily mean ergot has no effect in reducing the actual ('realized') total seed output of *Spartina* in Poole Harbour from what it would have been in the absence of ergot infection ('potential' seed output) because ergot may not infect inflorescences at random. If ergot preferentially infects inflorescences that have the highest potential seed output, there may be no detectable difference in seed production between infected and uninfected inflorescences, but realized seed production from these inflorescences would be lower than the potential output.

Bacon and Luttrell (1981) used radioactive carbon dioxide to demonstrate that ergots developing on rye plants diverted resources from developing seed. Raybould *et al.* (1998) found that heavy ergot infection of *Spartina* was associated with lower mean seed weight. Therefore, high amounts of ergot may reduce seed quality as well as seed number. Nevertheless seed set is sporadic in *Spartina* and seedling recruitment is rare in mature marshes such as those in Poole Harbour. Therefore, reduction in seed set may have little effect on *Spartina* fitness, except in the long term when release of genetic variation through sexual reproduction may be beneficial (Gray *et al.*, 1991).

A more immediately important effect of ergot may be diversion of resources from vegetative reproduction of the host. The mean weight of an ergot on a particular inflorescence was significantly negatively correlated with the number of ergots on the inflorescence in all years analysed (1986, 1988, 1992 and 1995) and variation in ergot number explained nearly three-quarters of the variation in mean ergot weight in 1995. This suggests resources are limiting ergot growth and so ergots may compete with the host for resources (Raybould *et al.*, 1998). However, the total weight of ergot per inflorescence is very small (>100 mg), and it seems unlikely that the growth of *Spartina* is greatly affected by diversion of resources to ergots. It is possible, however, that ergots may influence vegetative reproduction in other ways, perhaps by the production of chemicals that affect growth.

Conclusions

There are many opinions about whether the spread of *Spartina* is beneficial or detrimental to saltmarshes and surrounding areas (Doody, 1990). One certainty is that *Spartina* is fascinating to evolutionary biologists and ecologists because it presents the opportunity to study the ecological effects of speciation. The spread of *Spartina* was a massive perturbation to the ecosystem of Poole Harbour. The events described in this chapter represent physical and biological adjustments to this perturbation. The interesting question now is whether the invasion/erosion sequence is cyclical, with *Spartina* reinvading its old habitats, or whether it will reach equilibrium, perhaps at a much lower area of *Spartina*. The appearance of ergot infection is especially interesting as it indicates evolutionary as well as ecological change may now be occurring in response to the arrival of *Spartina*.

Acknowledgements

I especially thank Professor Alan Gray for many valuable discussions about Poole Harbour saltmarshes and Steve Pearce, the Hydrographic Surveyor of Poole Harbour

Commissioners, for the 1995 hydrographical data and much help in locating earlier data. I also thank Rebecca Mogg, Karen Jeal, Selwyn McGrorty, John Goss-Custard and Ralph Clarke, and R. N. Appleton, the Harbour Engineer, for permission to use the hydrographical data.

The editors are grateful to Judy Lindsay, Director of the Dorset County Museum for permission to reproduce this article from an original published in the *Proceedings of the Dorset Natural History and Archaeological Society.*

References

Bacon, C. W. and Luttrell, E. S. (1981) Competition between ergots of *Claviceps purpurea* on rye seed for photosynthates. *Phytopathology*, **72**: 1332–1336.

Bird, E. C. F. and Ranwell, D. S. (1964) *Spartina* marshes in southern England. IV. The physiography of Poole Harbour, Dorset. *Journal of Ecology*, **52**: 355–366.

Burd, F. (1989) *The Saltmarsh Survey of Great Britain.* Peterborough: Nature Conservancy Council.

Cayford, J. T. and Water, R. J. (1996) Population estimates for waders Charadrii wintering in Great Britain, 1987/88–1991/92. *Biological Conservation*, **77**: 7–17.

Davis, P. and Moss, D. (1984) *Spartina* and waders – the Dyfi Estuary. pp. 37–40. In: *Spartina anglica in Great Britain.* Doody, P. (ed.). Peterborough: Nature Conservancy Council.

Doody, P. (1990) *Spartina* friend or foe? A conservation viewpoint. pp. 77–79. In: *Spartina anglica – A Research Review.* Gray A. J. and Benham P. E. M. (eds). London: HMSO.

Eleuterius, L. N. and Meyers, S. P. (1974) *Claviceps purpurea* on *Spartina* in coastal marshes. *Mycologia*, **66**: 978–986.

Ferris, C., King, R. A. and Gray, A. J. (1997) Molecular evidence for the parentage in the hybrid origin of *Spartina anglica. Molecular Ecology*, **6**: 185–187.

Gessner, R. V. (1978) *Spartina alterniflora* seed fungi. *Canadian Journal of Botany*, **56**: 2942–2947.

Goodman, P .J. and Williams, W. T. (1961) Investigations into 'die-back' in *Spartina townsendii* agg. III. Physiological correlates of 'die-back'. *Journal of Ecology*, **49**: 391–398.

Goss-Custard, J. D. and Moser, M. E. (1988) Rates of change in the numbers of dunlin, *Calidris alpina*, wintering in British estuaries in relation to the spread of *Spartina anglica. Journal of Applied Ecology,* **25**: 95–109.

Gray, A. J. (1986a) *Poole Harbour: Ecological Sensitivity Analysis of the Shoreline.* Huntingdon: Institute of Terrestrial Ecology.

Gray, A. J. (1986b) Do invading species have definable genetic characteristics? *Philosophical Transactions of the Royal Society. Series B*, **314**: 655–674.

Gray, A. J., Drury, M, G. and Raybould, A. F. (1990) *Spartina* and the ergot fungus *Claviceps purpurea* – a singular contest? pp. 63–79. In: *Pests, Pathogens and Plant Communities.* Burdon J. J. and Leather S. R. (eds). Oxford: Blackwell Scientific Publications.

Gray, A. J., Marshall, D. F. and Raybould, A. F. (1991) A century of evolution in *Spartina anglica. Advances in Ecological Research*, **21**: 1–62.

Gray, A. J. and Pearson, J. M. (1984) *Spartina* marshes in Poole Harbour, Dorset, with special reference to Holes Bay. pp. 11–14. In: *Spartina anglica in Great Britain.* Doody, P. (ed.). Peterborough: Nature Conservancy Council.

Green, F. H. W. (1940) *Poole Harbour. A Hydrographic Survey.* London: Geographical Publications.

Hubbard, J. C. E. (1965) *Spartina* marshes in southern England. VI. Pattern of invasion in Poole Harbour. *Journal of Ecology*, **53**: 799–813.

Hubbard, J. C. E. and Stebbings, R. E. (1967) Distribution, dates of origin, and acreage of *Spartina townsendii* (*s.l.*) marshes in Great Britain. *Proceedings of the Botanical Society of the British Isles*, **7**: 1–7.

Hubbard J. C. E. and Stebbings, R. E. (1968) *Spartina* marshes in southern England. VII. Stratigraphy of the Keysworth Marsh, Poole Harbour. *Journal of Ecology*, **56**: 707–722.

Jenkinson, J. G. (1958) Ergot infection of grasses in the south-west of England. *Plant Pathology*, **7**: 81–85.

Lambert, J. M. (1964) The *Spartina* story. *Nature*, **204**: 1136–1138.

McGrorty, S. and Goss-Custard, J. D. (1987) *A Review of the Rehabilitation of Areas Cleared of Spartina.* NCC/NERC Contract Report HF3/08/21(2). Huntingdon: Institute of Terrestrial Ecology.

MAFF (1974) *Ergot of Cereals and Grasses. Ministry of Agriculture, Fisheries and Food Agricultural Development and Advisory Service Advisory Leaflet,* No. 548.

Marchant, C. J. (1967) Evolution of *Spartina* (Gramineae). I. The history and morphology of the genus in Britain. *Botanical Journal of the Linnean Society*, **60**: 1–24.

Marchant, C. J. (1968) Evolution of *Spartina* (Gramineae). II. Chromosomes, basic relationships and the problem of *S.* x *townsendii* agg. *Botanical Journal of the Linnean Society*, **60**: 381–409.

May, V. J. (1969) Reclamation and shoreline change in Poole Harbour, Dorset. *Proceedings of the Dorset Natural History and Archaeological Society*, **90**: 141–154.

Millard, A. V. and Evans, P. R. (1984) Colonization of mudflats by *Spartina anglica*; some effects on invertebrate and shorebird populations. pp. 41–48. In: *Spartina anglica in Great Britain.* Doody, P. (ed.). Peterborough: Nature Conservancy Council.

Oliver, F. W. (1920) *Spartina* problems. *Annals of Applied Biology*, **7**: 25–39.

Oliver, F. W. (1925) *Spartina townsendii*; its mode of establishment, economic uses, and taxonomic status. *Journal of Ecology*, **13**: 74–91.

Poole Harbour Commissioners (1982) *Poole Harbour Master Plan Study Parts I and II.*

Prater, A. J. (1981) *Estuary Birds of Britain and Ireland.* Calton: T. & A. D. Poyser.

Ranwell, D. S. (1964) *Spartina* marshes in southern England. II. Rate and seasonal pattern of sediment accretion. *Journal of Ecology*, **52**: 79–94.

Ranwell, D. S. (1967) World resources of *Spartina townsendii* (*s.l.*) and economic use of *Spartina* marshland. *Journal of Ecology*, **52**: 95–105.

Ranwell, D. S., Bird, E. C .F., Hubbard, J. C. E. and Stebbings, R. E. (1964) *Spartina* marshes in southern England. V. Tidal submergence and chlorinity in Poole Harbour. *Journal of Ecology*, **52**: 627–641.

Raybould, A. F., Gray, A. J. and Clarke, R. T. (1998) The long-term epidemic of *Claviceps purpurea* on *Spartina anglica* in Poole Harbour: pattern of infection, effects on seed production and the role of *Fusarium heterosporum*. *New Phytologist*, **138**: 497–505.

Raybould, A. F., Gray, A. J., Lawrence, M. J. and Marshall, D. F. (1991a) The evolution of *Spartina anglica* C.E.Hubbard (Gramineae): origin and genetic variability. *Biological Journal of the Linnean Society*, **43**: 111–126.

Raybould, A. F., Gray, A. J., Lawrence, M. J. and Marshall, D. F. (1991b) The evolution of *Spartina anglica* C. E. Hubbard (Gramineae): genetic variation and status of the parental species in Britain. *Biological Journal of the Linnean Society*, **44**: 369–380.

Schloten, M. T. C. and Rozema, J. (1990) The competitive ability of *Spartina* on Dutch saltmarshes. pp. 39–47. In: *Spartina anglica – A Research Review*. Gray, A. J. and Benham, P. E. M. (eds). London: HSMO.

Stapf, O. (1913) Towndsend's grass or rice grass. *Proceedings of the Bournemouth Natural Science Society*, **5**: 76–82.

Townsend, F. (1883) *Flora of Hampshire*. London: Reeve.

Tubbs, C. R., Tubbs, J. M. and Kirby, J. S. (1992) Dunlin *Calidris alpina alpina* in the Solent, southern England. *Biological Conservation*, **60**: 15–24.

7. Macro-invertebrate Fauna in the Intertidal Mudflats

Richard Caldow[1], Selwyn McGrorty[1], Andrew West[1], Sara E. A. le V. dit Durell[1], Richard Stillman[1] and Sheila Anderson[2]

[1]Centre for Ecology and Hydrology, CEH Dorset, Winfrith Technology Centre, Dorchester, Dorset DT2 8ZD

[2]Natural Environment Research Council, Polaris House, North Star Avenue, Swindon, Wiltshire SN2 1EU

Poole Harbour is a Special Protection Area (SPA) because of its importance to breeding and wintering waders and wildfowl. English Nature commissioned the Centre for Ecology and Hydrology to survey the intertidal food resources of these birds. Macro-invertebrate populations were sampled at 80 stations across the harbour. *Cirratulis filiformis* and *Tubificoides benedini* were by far the most numerically abundant species. *Nereis virens* and *Cirratulis filiformis* were the most abundant species in terms of biomass. Variation in the numerical or biomass density of macro-invertebrates across the harbour does not explain the birds' distribution. In terms of numerical densities, the macro-invertebrate community in Poole Harbour is similar to that on the Exe estuary. In terms of biomass, however, Poole Harbour is dominated by worms, whereas the Exe estuary is dominated by molluscs. A review of historical surveys of macro-invertebrates in Poole Harbour indicates that change rather than stability is the norm.

Introduction

Poole Harbour supports populations of waders and wildfowl that are of local, national and, in some cases, international importance (Pickess and Underhill-Day, 2002). In recognition of this, the harbour has been designated under national and international legislation. English Nature is required under European Union regulations to ensure that the interest features of the harbour, as a Special Protection Area (SPA), are maintained in "favourable condition". Accordingly, English Nature commissioned the Centre for Ecology and Hydrology to survey the macro-invertebrate abundance and biomass in the intertidal sediments of Poole Harbour. This will serve as a baseline against which to compare future surveys to detect any significant change in prey abundance and hence the "favourable condition" of the interest features of the harbour.

Methods

Survey plan

A 500 m x 500 m grid, based on the Ordnance Survey grid, superimposed on a map of the harbour, resulted in 80 intersections falling between Mean High Water and Mean Low Water. These sampling stations were spread evenly across the intertidal areas of Poole Harbour to which birds regularly gain access.

Field methods

The survey was conducted during September and October 2002. Sampling stations were visited either at low water on spring tides (on foot or by hovercraft) or at high water on neap tides (by boat). Three sampling methods were employed at each sampling station.

(i) A single 10.6 cm diameter x 30 cm deep sediment core was removed with a steel pipe. Each core was sieved through a 0.5 mm nylon mesh. Sieve contents were fixed in 4% formalin and, after a number of days, were washed in freshwater and preserved in IMS.

(ii) A randomly chosen area of 0.25 m^2 was 'dredged' using a hand-net with a mesh-size of 2 mm. The larger, near-surface dwelling organisms were collected. These samples were frozen.

(iii) A randomly assigned 1 m^2 area of the mud surface was inspected and the numbers of *Arenicola marina* casts counted.

For each of the key species, a number of additional individuals spanning the range of sizes present in the harbour were collected. Single live specimens were placed in individual plastic bags and frozen.

Sample processing

Benthic invertebrates

All macro-invertebrates in each core and net sample were counted and identified to the lowest necessary level of taxonomic detail (i.e. species level for all except small worms). The length of all individuals of the key species was measured.

The numerical densities were derived from the core samples for all species except *Carcinus maenas*, *Crangon crangon*, *Tapes philippinarum*, *Crepidula fornicata*, *Macoma balthica*, all *Littorina* spp., *Gibbula umbilicalis* and *Hinia reticulata*. The numerical densities of these species were derived from the netted samples. The overall numerical density of *Cerastoderma edule* at each sampling station was calculated by combining estimates of the density of individuals <6 mm from the core and of individuals >6 mm from the net sample.

Biomass-length relationships

For each species (excepting small worms), each specimen was defrosted, measured and then processed according to standard laboratory procedures to yield its biomass in terms of ash-free dry mass (AFDM). In the case of molluscs that are typically opened by birds to remove the flesh, i.e. bivalves, only the flesh of each individual was processed. In the case of molluscs and crustaceans that are typically eaten whole by birds, e.g. *Hydrobia* spp. and *Carcinus maenas*, each animal was processed intact.

Calculation of biomass densities

For each species, the raw data relating the AFDM of an individual to its length were transformed ($\log_e$) and a linear regression model fitted to the data. Species-specific regression equations were used to predict the AFDM of an individual within each millimetre size class across the full size range for that species. For each species, the numerical density of individuals within each millimetre size class at each station was multiplied by the relevant predicted value of AFDM to yield the biomass density of that size class at that station. Biomass densities were summed across size classes to yield the total biomass density of a species at each station.

In the case of *Arenicola marina*, for which only the density of casts was assessed, the biomass density at each station was calculated by multiplication of cast density by the average AFDM of all the individuals processed to derive the AFDM-length relationship for that species.

In the case of many of the small tube-dwelling worms, and other small worms, that were not assigned to size classes, the biomass density at a station was calculated by multiplying the numerical density by the average AFDM of a small worm that was derived from ashing 100 such small worms *en masse*.

Biomass densities were derived for all species whose average numerical density across the harbour exceeded 2 individuals m^{-2}.

The effects of invertebrate distribution on bird distribution

Each sampling station was allocated to the appropriate Wetland Bird Survey (WeBS) count sector (Map 3 of Pickess and Underhill-Day, 2002) and the average overall numerical and biomass density of macro-invertebrates was calculated for each. The bird count data presented in Pickess and Underhill-Day (2002, p. 141) was re-worked excluding waterfowl. For each WeBS count sector, the sum of the harbour-wide percentages of the eight wader species (plus Shelduck) held by that sector was calculated and then expressed as a percentage of 900, i.e. the harbour-wide, across species sum of these percentages. These figures were regressed against the area of mud exposed at low water on Spring tides within each sector (see Pickess and Underhill-Day, 2002, pp. 6–7). The extent to which variation in the numerical or biomass density of invertebrates between sectors explained the residual variation in bird usage was explored.

Results

Species presence and distribution

Sixty-one kinds of macro-invertebrates were identified (Table 1). Species-richness was fairly even around the harbour, although the upper Wareham Channel and the bays on its north-east side were the least rich areas (Figure 1). The numerical density of macro-invertebrates was, however, high in the upper reaches of the Wareham channel and several other sheltered areas (Figure 2). The biomass density of macro-invertebrates was quite patchy around the harbour (Figure 3). Some of the quiet backwaters on the southern shore (Middlebere Lake, Wych Lake and Brands Bay) were poor in this respect. Hotspots of biomass density occurred around the harbour, with the greatest concentration being around Baiter and Parkstone Bay (Figure 3).

Only a very few species occurred throughout the harbour, the most cosmopolitan being *Tubificoides benedini* and *Cerastoderma edule. Cirratulus filiformis*, the most numerous species, was absent from the harbour's more seaward bays as was *Hediste diversicolor. Hediste diversicolor* was replaced at the seaward end of the harbour by *Nereis virens. Arenicola marina* was particularly abundant immediately adjacent to the harbour mouth. *Cyathura carinata*, one of the most abundant crustaceans, was restricted to sheltered mudflats far from the harbour mouth. In contrast, *Abra tenuis*, the most abundant bivalve mollusc, was distributed throughout the harbour, being absent only from those areas immediately opposite the harbour mouth and around the bays on the north-east shore of the harbour. *Hydrobia* spp. were the third most abundant species overall and were widely distributed but especially abundant in Lytchett Bay. Detail on the distribution of the species in Poole Harbour is available in Thomas *et al.* (2004).

Numerical and biomass densities

The average harbour-wide numerical and biomass densities of macro-invertebrates are presented in Table 1. Only two species (*Cirratulus filiformis* and *Tubificoides benedini*) occurred at average densities of over 1000 individuals m^{-2}. When the species were ranked according to their average biomass density, several of the rarer but larger molluscs (*Mya arenaria, Cerastoderma edule, Tapes philippinarum* and *Scrobicularia plana*) became considerably more important. Nonetheless, three of the four top-ranked species in terms of biomass are worms (Tables 1 and 3).

Comparison between Poole Harbour and the Exe estuary

The macro-invertebrates of the Exe estuary were surveyed in autumn 2001 (Durell *et al.* 2005). Here, 59 kinds of invertebrates were identified (Table 2). The top three species in terms of numerical density were the same as in Poole Harbour (Tables 1–3).When the species were ranked according to their average biomass density, several mollusc species became markedly more important. In contrast to Poole Harbour, however, three of the four top-ranked species in terms of biomass, were bivalve molluscs, the mussel *Mytilus edulis* being by far the most dominant species (Tables 2 and 3).

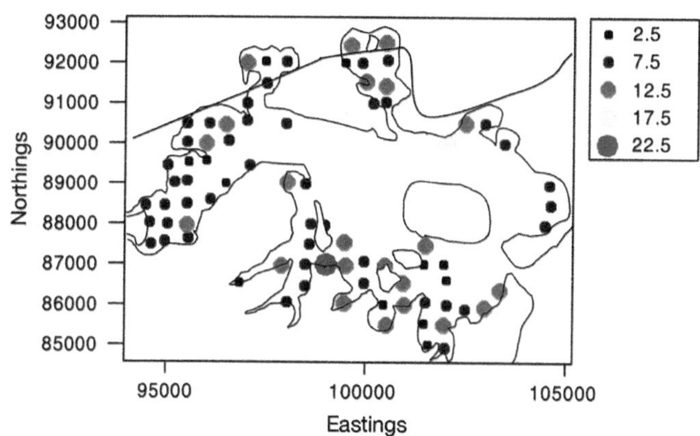

Figure 1 An illustrative sketch of Poole Harbour showing the location of the 80 sampling stations and the number of species of macro-invertebrate found at each. In this figure, as in Figures 2 and 3, the values in the key denote the mid-points of bands into which stations have been grouped. (i.e. in this case 2.5 = 1–5 spp., 7.5 = 5–10, 12.5 = 10–15, 17.5 = 15–20, 22.5 = 20–25).

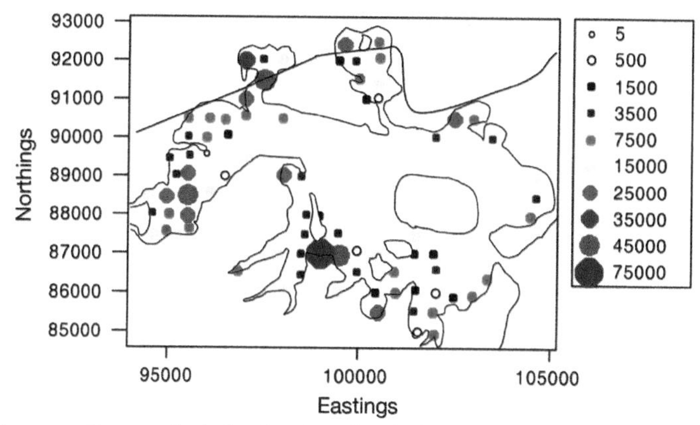

Figure 2 The overall numerical density (n m^{-2}) of all macro-invertebrates at each of the 80 sampling stations.

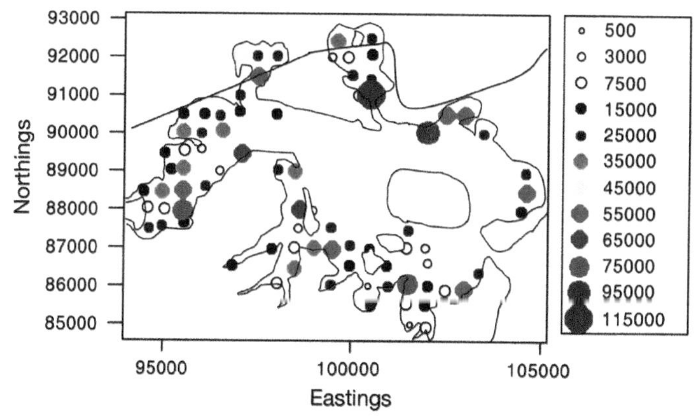

Figure 3 The overall biomass density (mg ash-free dry mass m^{-2}) of macro-invertebrates at each of the 80 sampling stations.

Table 1 The harbour-wide average numerical density of all 'species' of macro-invertebrate (in rank order) and the average biomass density of those macro-invertebrates whose numerical density exceeded 2 m^{-2} in Poole Harbour (also in rank order)

Species	Average number m^{-2}	Species	Average mg ash-free dry mass m^{-2}
Cirratulus filiformis	3819.6	*Nereis virens*	4567
Tubificoides benedini	1818.5	*Cirratulus filiformis*	4545
Hydrobia spp.	756.3	*Hediste diversicolor*	2453
Microdeutopus gryllotalpa	635.9	*Mya arenaria*	2323
Hediste diversicolor	614.7	*Tubificoides benedini*	2164
Malacoceros fuliginosus	422.0	*Cerastoderma edule*	2155
Corophium volutator	373.9	*Arenicola marina*	1176
Cyathura carinata	355.5	*Tapes philippinarum*	1155
Gammarus locusta	277.6	*Scrobicularia plana*	1093
Scoloplos armiger	263.4	*Nepthys hombergii*	552
Abra tenuis	254.9	*Littorina* spp.	537
Spionid spp.	151.5	*Malacoceros fuliginosus*	502
Anemones (unident)	114.7	*Hydrobia* spp.	449
Capitella capitata	113.3	*Scoloplos armiger*	313
Nereis virens	80.7	*Abra tenuis*	301
Eteone longa	60.9	*Cyathura carinata*	252
Chironomid larvae	51.0	*Carcinus maenas*	213
Nepthys hombergii	46.7	Spionid spp.	180
Urothoe poseidonis	42.5	*Corophium volutator*	148
Neomysis integer	35.4	*Capitella capitata*	135
Ampharete grubei	34.0	*Eteone longa*	72
Cerastoderma edule	30.2	*Neomysis integer*	54
Polycirrus caliendrum	29.7	*Microdeutopus gryllotalpa*	48
Corophium arenarium	28.3	*Ampharete grubei*	40
Scolelepis foliosa	19.8	*Gammarus locusta*	39
Mesopodopsis slabberi	15.6	*Polycirrus caliendrum*	35
Nemerteans	12.7	*Scolelepis foliosa*	24
Scolelepis squamata	11.3	*Corophium arenarium*	20
Pygospio elegans	11.3	*Mesopodopsis slabberi*	16
Anaitides mucosa	9.9	Nemerteans	15
Cirriformia tentaculata	9.9	*Praunus flexuosus*	15
Heteromastus filiformis	9.9	*Scolelepis squamata*	13
Nematodes	9.9	*Pygospio elegans*	13
Praunus flexuosus	9.9	*Anaitides mucosa*	12
Mya arenaria	9.9	*Cirriformia tentaculata*	12
Scrobicularia plana	8.5	*Heteromastus filiformis*	12
Gammaropsis palmata	7.1	Nematodes	12
Tapes philippinarum	4.6	*Urothoe poseidonis*	7
Idotea neglecta	4.2	*Palaemon longirostrus*	5
Idotea pelagica	4.2	*Bathyporeia sarsi*	3
Chironomid pupae	4.2	*Idotea neglecta*	2
Littorina spp.	4.0	*Harmothoe* spp.	2
Arenicola marina	3.3	*Glycera tridactyla*	2
Bathyporeia sarsi	2.8		
Palaemon longirostrus	2.8		
Carcinus maenas	2.1		

Table 1 ***cont.***

Species	Average number m^{-2}	Species	Average mg ash free dry mass m^{-2}
Harmothoe spp.	1.4		
Glycera tridactyla	1.4		
Idotea balthica	1.4		
Talitrus saltator	1.4		
Palaemon serratus	1.4		
Macoma balthica	1.4		
Haminoea navicula	1.4		
Crangon crangon	1.2		
Crepidula fornicata	0.9		
Ascidians (unident)	0.4		
Gibbula umbilicalis	0.4		
Hinia reticulata	0.2		
Mysidae spp.	0.1		

Table 2 The estuary-wide average numerical density (in rank order) and the average biomass density (also in rank order) of all 'species' of macro-invertebrate on the Exe estuary in autumn 2001

Species	Average number m^{-2}	Species	Average mg ash-free dry mass m^{-2}
Hydrobia spp.	2925.8	*Mytilus edulis*	21628
Tubificoides benedeni	2509.8	*Scrobicularia plana*	9993
Cirratulid spp.	2103.5	*Hediste diversicolor*	6911
Spio spp.	1632.5	*Cerastoderma edule*	3892
Hediste diversicolor	944.1	*Littorina* spp.	3724
Pygospio elegans	794.6	*Hydrobia* spp.	1760
Heteromastus filiformis	279.7	*Tubificoides benedeni*	863
Cyathura carinata	245.2	*Cirratulid* spp.	724
Capitella capitata	193.7	*Spio* spp.	562
Scrobicularia plana	151.1	*Nepthys hombergii*	457
Ampharete acutifrons	129.4	*Lanice conchilega*	432
Cerastoderma edule	118.2	*Carcinus maenus*	329
Gammarus locusta	112.1	*Crangon crangon*	316
Corophium arenarium	110.1	*Pygospio elegans*	273
Tubifex spp.	109.3	*Cyathura carinata*	246
Corophium volutator	94.8	*Corophium volutator*	110
Malacoceros fuliginosus	88.0	*Arenicola marina*	106
Eteone longa	85.2	*Heteromastus filiformis*	96
Mytilus edulis	72.7	*Capitella capitata*	67
Littorina spp.	69.9	*Malacoceros fuliginosus*	59
Scolelepis squamata	49.0	*Gammarus locusta*	50
Nepthys hombergii	47.0	*Tubifex* spp.	38
Crangon crangon	44.2	*Ampharete acutifrons*	37
Neomysis integer	42.2	*Scolelepis squamata*	33
Nematodes	39.8	*Eteone longa*	29
Lanice conchilega	29.3	*Corophium arenarium*	26

Table 2 *cont.*

Species	Average number m^{-2}	Species	Average mg ash-free dry mass m^{-2}
Carcinus maenus	26.5	*Anaitides mucosa*	24
Bathyporeia sarsi	20.9	*Ophelia bicornis*	23
Urothoe poseidonis	16.1	*Neomysis integer*	22
Anaitides mucosa	13.7	*Bathyporeia sarsi*	21
Praunus flexuosus	11.7	*Urothoe poseidonis*	8
Dipteran larva	9.6	*Scoloplos armiger*	3
Angulus tenuis	9.6		
Nemerteans	5.6		
Sphaeroma serratum	5.6		
Mya arenaria	5.6		
Scoloplos armiger	5.2		
Ophelia bicornis	5.2		
Arenicola marina	4.0		
Bathyporeia pelagica	3.2		
Gibbula umbilicalis	3.2		
Macoma balthica	3.2		
Idotea pelagica	2.8		
Jaera albifrons	2.8		
Idotea chelipes	2.4		
Chironimid larva	2.4		
Glycera tridactyla	2.0		
Kefersteinia cirrata	2.0		
Tapes decussatus	1.6		
Melita palmata	1.2		
Eurydice pulchra	1.2		
Eteone viridis	0.8		
Euclymene lumbricoides	0.8		
Harmothoe spp.	0.8		
Tanaid sp.	0.8		
Crepidula fornicata	0.8		
Abra alba	0.8		
Lepidochitona cinereus	0.8		
Nematonereis unicornis	0.8		

Source: Durell *et al.* (2005).

Historical perspective

The macro-invertebrate populations of the intertidal flats of Poole Harbour have been surveyed several times (Table 4). In the early 1970s, the Nature Conservancy surveyed the entire harbour. In 1987, several independent studies combined to yield an extensive survey. *Scrobicularia plana* and *Macoma balthica* are considerably scarcer now than they have been over the last 30 years (Table 5). In contrast, other bivalves, notably *Cerastoderma edule* and in particular *Abra tenuis* have become more abundant. *Hydrobia* spp. are also more abundant now as are errant polychaetes. Sedentary polychaetes may also have increased since the 1970s. In contrast, *Corophium volutator* seems to be far less abundant now than it has been in the past, but increases in other crustaceans (*Microdeutopus gryllotalpa* and

Table 3 Summary of key statistics and characteristics of the intertidal macro-invertebrate communities in Poole Harbour and the Exe estuary

	Poole Harbour	Exe estuary	Species in common
Total 'species'	61	59	38
Molluscs	15	13	8
Polychaete worms	20	22	14
Crustaceans	20	17	11
Numerical density Top three species (n m^{-2})	*Cirratulus filiformis* *Tubificoides benedini* *Hydrobia* spp.	*Hydrobia* spp. *Tubificoides benedini* *Cirratulus filiformis*	*Hydrobia* spp. *Tubificoides benedini* *Cirratulus filiformis*
N spp. >1000 m^{-2}	2	4	*Tubificoides benedini* *Cirratulus filiformis*
N spp. >100 m^{-2} and < 1000 m^{-2}	12	11	*Hediste diversicolor* *Cyathura carinata* *Gammarus locusta* *Capitella capitata*
N spp. >10 m^{-2} and < 100 m^{-2}	15	16	*Eteone longa* *Nepthys hombergii* *Urothoe poseidonis* *Neomysis integer* *Scolelepis squamata*
N spp. < 10 m^{-2}	30	28	*Mya arenaria* *Idotea pelagica* Chironomids *Arenicola marina* *Harmothoe* spp. *Glycera tridactyla* *Macoma balthica* *Crepidula fornicata* *Gibbula umbilicalis*
Overall numerical density (n m^{-2})	10603	13195	
Biomass density Top species (mg AFDM m^{-2})	*Nereis virens* *Cirratulus filiformis* *Hediste diversicolor* *Mya arenaria*	*Mytilus edulis* *Scobicularia plana* *Hediste diversicolor* *Cerastoderma edule*	*Hediste diversicolor*
Biomass density N spp. > 1000 mg AFDM m^{-2}	9	6	*Hediste diversicolor* *Cerastoderma edule* *Scrobicularia plana*
Biomass density N spp. > 100 mg AFDM m^{-2} and < 1000 mg AFDM m^{-2}	11	11	*Nepthys hombergii* *Cyathura carinata* *Carcinus maenas* Spionid sp. *Corophium volutator*
Overall biomass density	25689	52862	
Dominant group in terms of biomass	Polychaete worms	Bivalve molluscs	

Source: Data for Exe estuary from Durell *et al.* (2005).

Table 4 Details of historical surveys of intertidal macro-invertebrates in Poole Harbour

Year	Location	Number of sampling stations	Numerical densities	Size classes	Source
1971	Various locations	8	yes	no	Arnold (1971)
1972	Whole harbour	189	yes	no	Anderson *et al.* (unpublished data)
1982–83	Parkstone Bay	75	yes	no	Harris (1983a)
1982–83	Holes Bay	?	?	?	Harris (1983b)
1985	Various locations	45	yes	no	Institute of Offshore Engineering (1986)
1987	Holes Bay	22	yes	no	Dyrynda (1989)
1987	Cleavel - Ower	41	yes	no	McGrorty *et al.* (1987)
1987	Whitley Lake, Brands Bay, Newton Bay, Keysworth	105	yes	(*Hediste* and *Nepthys* only)	Warwick *et al.* (1989)
1991	Holes Bay	31	yes	no	Environment Agency (unpublished report)
1991	Wych Lake and Channel	15	yes	no	Jensen *et al.* (1991)
1999	Arne Bay, Brands Bay, Newton Bay	36	yes	no	J. Gill (unpublished data)
2002	Whole harbour	80	yes	yes	This study
2002	Holes Bay	10	yes	yes	Caldow *et al.* (2003)

Cyathura carinata) have more than offset this. The overall number of macro-invertebrates has been relatively stable over the last 20 years but (subject to the effect of differing sieve mesh sizes) has increased since the 1970s. This is largely due to a marked increase in the numbers of sedentary polychaetes and other worms (e.g. oligochaetes).

Holes Bay has been surveyed in each of the last four decades. Within Holes Bay, very similar patterns to those seen across the harbour as a whole are apparent (Table 6). *Scrobicularia plana* and *Macoma balthica* have been virtually absent since the early 1970s. In contrast, *Cerastoderma edule* and particularly *Abra tenuis* have increased markedly over the same time. *Hydrobia* spp. are also much more abundant now than in the past. Both errant and sedentary polychaetes have fluctuated dramatically between decades, although not in synchrony. Consequently, the total density of all worms has been fairly similar in all decades except the 1980s. *Corophium volutator* abundance has also fluctuated widely. Other crustaceans seem to have increased since the 1980s, the numbers being dominated by Ostracoda and *Cyathura carinata* in the 1990s and by *C. carinata*, *Microdeutopus gryllotalpa* and *Gammarus locusta* in 2002. The total count of

Table 5 Comparison of the average numerical densities of key species and groups of macro-invertebrates on intertidal flats in Poole Harbour over the last 30 years

	Source		
	Anderson *et al.* (unpublished data)	**Dyrynda (1989); McGrorty *et al.* (1987); Warwick *et al.* (1987)**	**Current study**
Year survey	**1972**	**1987**	**2002**
Scrobicularia plana	53	239	8
Macoma balthica	42	10	1
Other bivalves	27	133	300
All bivalves	120	383	310
Hydrobia spp.	214	135	756
Errant polychaetes	376	478	814
Sedentary polychaetes	2023	6570	4909
Other worms	6	1884	1841
All worms	2398	8931	7564
Corophium spp.	1540	2882	374
Other crustaceans	56	177	1430
All crustaceans	1577	3059	1804
All	4309	12508	10434
Number of samples	189	168	80

all crustaceans has fluctuated widely. In spite of these fluctuations, the total invertebrate numbers have been fairly consistent over the four decades with the notable exception of the 1980s when numbers of all groups were much reduced.

Bird and invertebrate associations

When the WeBS count data were restricted to exclude waterfowl (except Shelduck) and to the 24 count sectors for which invertebrate data were available, there was a highly significant association between bird usage and the area of mud within a sector ($r^2 = 63.8\%$, d.f. = 23, $P < 0.001$) (Figure 4). Across the harbour there was a significant but imperfect correlation between the numerical and biomass density of invertebrates ($r = 0.32$, $n = 80$, $P < 0.01$) (Figure 5). The residual scatter around the birds vs area of mud regression line was not, however, associated with either the variation in the numerical ($P = 0.733$) or biomass ($P = 0.082$) density of macro-invertebrates between the sectors (Figure 6).

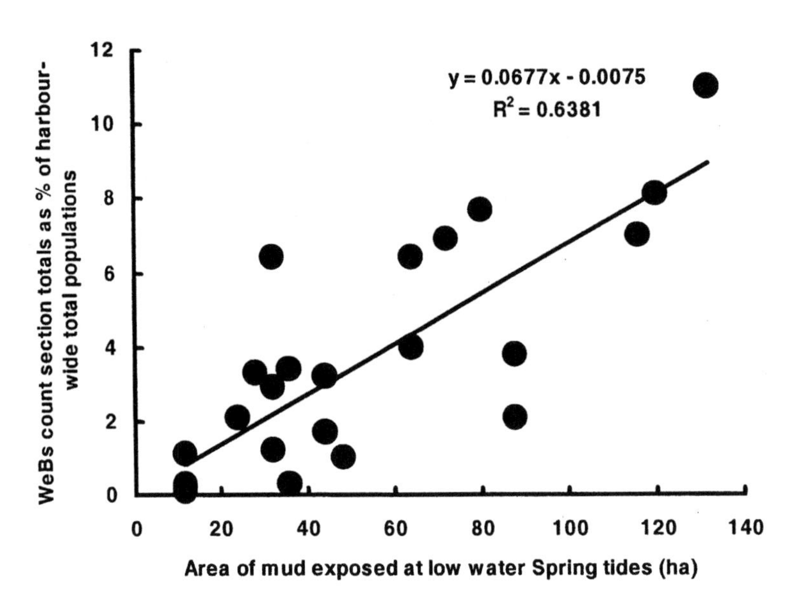

Figure 4 The relationship between the percentage of the harbour-wide populations of all waders and Shelduck within each of the WeBS count sectors at low water and the area of intertidal flats exposed within that sector at low water (data taken from Pickess and Underhill-Day, 2002).

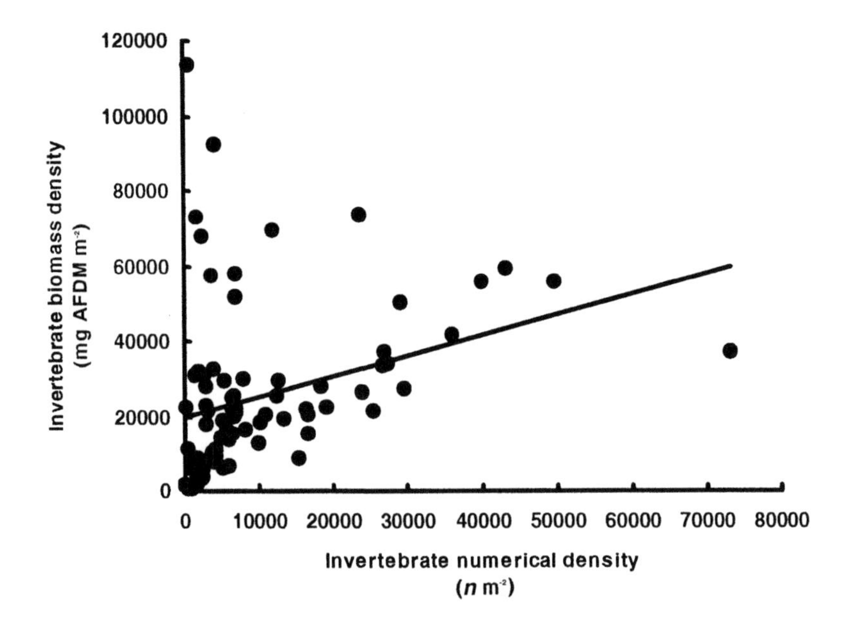

Figure 5 The relationship between the overall biomass density and overall numerical density of invertebrates at each sampling station. The solid line is the line of least squares best fit.

a)

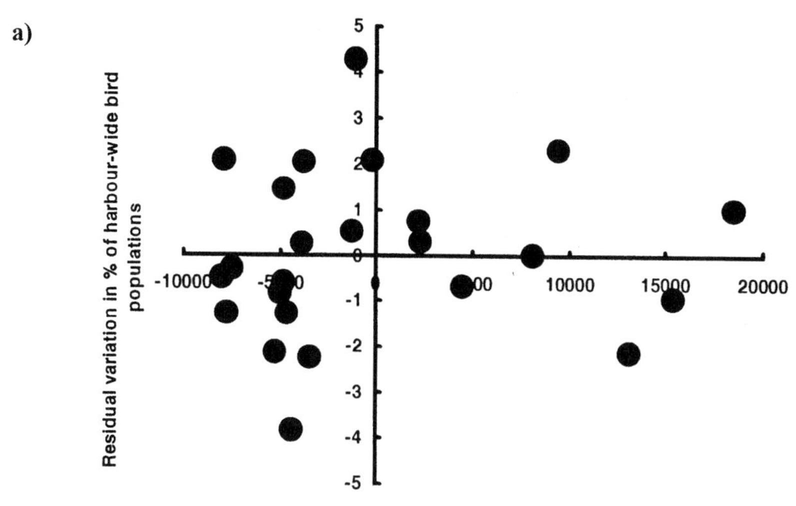

b)

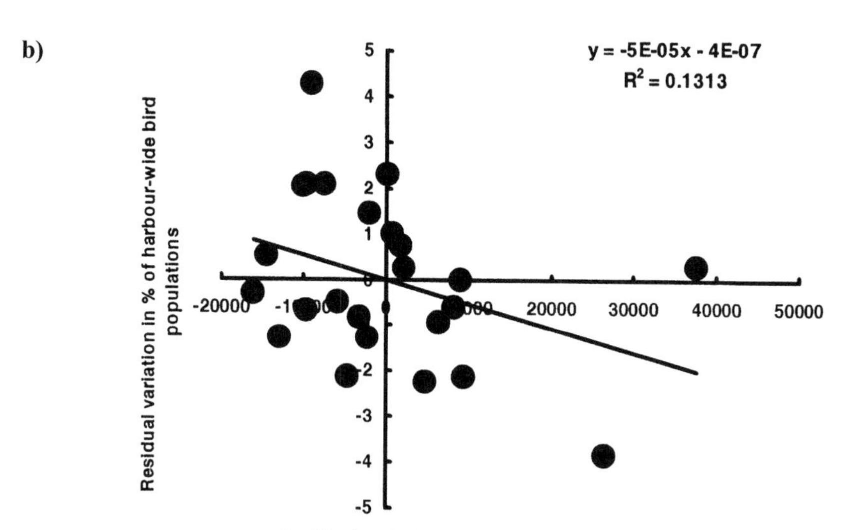

Figure 6 Plots of the residual variation in the percentage of the harbour-wide populations of all waders and Shelduck within each of the WeBS count sectors at low water that is not explained by the area of intertidal flats exposed within that sector at low water against: a) the residual variation in the macro-invertebrate numerical density (n m^{-2}) within a WeBS count sector that is also not explained by the area of intertidal flats exposed within that sector at low water; and b) the residual variation in the macro-invertebrate biomass density (mg AFDM m^{-2}) within a WeBS count sector that is not explained by the area of intertidal flats exposed within that sector at low water. These plots represent the relationships between the bird usage of a sector and its macro-invertebrate population, controlling for the partial effect of mudflat area on both.

Table 6 Comparison of the average numerical densities of key species and groups of macro-invertebrates on intertidal flats of Holes Bay, Poole Harbour over the last 30 years

	Source			
	Anderson *et al.* (unpublished data)	Dyrynda (1989)	Environment Agency (unpublished report)	Current study
Year survey	1972	1987	1991	2002
Scrobicularia plana	32	0	4	6
Macoma balthica	77	1	1	0
Other bivalves	43	4	780	294
All bivalves	146	5	785	300
Hydrobia spp.	31	0	253	1013
Errant polychaetes	619	123	2750	992
Sedentary polychaetes	5008	177	582	3491
Other worms	0	205	576	610
All worms	5627	506	3908	5093
Corophium spp.	2457	50	623	13
Other crustaceans	14	3	513	835
All crustaceans	2265	53	1135	848
All	8069	563	6081	7254
Number of samples	21	22	31	19

Discussion

The current macro-invertebrate community

The richness of the macro-invertebrate community is fairly uniform across the whole harbour. However, very few species were ubiquitous. Only *Tubificoides benedini* and *Cerastoderma edule* occur throughout the harbour. Amongst the other species, there is a wide range of distribution patterns. Some species are restricted to the areas furthest from the harbour mouth, e.g. *Cirratulus filiformis* and *Hediste diversicolor*, others to areas near the harbour mouth, e.g. *Arenicola marina*. Some species, e.g. *Cyathura carinata,* appear to favour the quietest backwaters. All of these species-specific distribution patterns will reflect the tolerance of each species to environmental factors that vary across the harbour. As a consequence of the varying species-specific distributions, both the overall numerical density and the overall biomass density are highly variable around the harbour. However, these two parameters are not perfectly correlated. The locations with the greatest numerical densities are generally in the quieter backwaters (upper reaches of the Wareham Channel, Lytchett Bay, around Fitzworth Point). There are, however, numerous other areas where the biomass density is high. The poorest areas of the harbour in terms of both numerical and biomass density are Brands Bay, the low-level sandflats opposite the harbour mouth and the north shore of the upper Wareham Channel.

Birds and invertebrates

Although every WeBS count sector of the harbour supports several species of waders or waterfowl, only a few sectors support over 5% of their combined populations (Pickess and Underhill-Day, 2002). When the bird data are restricted to the eight species of wader plus Shelduck, there is a strong positive association between the extent of mud within a sector and the percentage of birds held by it. There is, however, considerable scatter around this relationship. Several areas are markedly under-used, e.g. Lytchett Bay, the south-eastern quarter of Holes Bay, Sandbanks and the eastern (outer) part of Brands Bay. In contrast, the western (inner) part of Brands Bay is over-used by the birds, as are the areas around Fitzworth, Wych Lake, Keysworth and Swineham. Such 'outliers' cannot be explained by either a shortage or an excess of food supplies. Three of the four most under-used sectors have relatively high invertebrate biomasses whereas all five of the most over-used sectors have relatively low invertebrate biomasses. Pickess and Underhill-Day (2002) speculated that human disturbance may be a factor driving birds' distribution in the harbour. Three of the under-used areas (Brands Bay (east), Sandbanks and the south-eastern quarter of Holes Bay) are all subject to heavy boat traffic. It is unclear at present why Lytchett Bay should be so under-utilized.

Further work will be required to identify the reasons underlying the variation around the bird usage vs area relationship. At the crudest level of analyses in which all birds and invertebrates are considered together, the food supply cannot explain the discrepancies. A more detailed analysis in which the distributions of individual bird species are analysed in relation to the abundance of only their preferred prey species may yield more powerful insights into this issue. However, the fact that all of the under-used sectors are not impoverished in terms of food, and all the over-used areas are not the best in terms of food supplies, raises a number of interesting issues.

The decision to survey the macro-invertebrates across the entire harbour is fully justified. The survey has revealed several areas that are currently under-exploited by the birds yet are not lacking in food. The harbour may be capable of supporting many more birds than at present. Birds displaced from currently favoured areas may be able to relocate to under-utilized parts of the harbour where there is as much or even more food than where they currently feed. Conversely, perhaps human disturbance or other factors, e.g. green algal cover, restrict the usage of certain areas and hence the populations of birds that the harbour currently supports.

Poole Harbour 2002 survey in context

Comparison with the Exe estuary

Poole Harbour and the Exe estuary in Devon are remarkably similar in terms of:

- the total number of macro-invertebrate species identified
- the balance in species numbers between molluscs, polychaete worms and crustaceans
- the most numerically abundant species
- the frequency distribution of species numerical abundance
- the overall numerical density.

However, this similarity disappears when the biomass of the macro-invertebrates is considered. Worms dominate the biomass of Poole Harbour whereas bivalve molluscs dominate the biomass on the Exe estuary. Although the frequency distribution of species' biomass densities is similar, the numbers of species in common between the two sites in the biomass rank orders is much reduced. The Exe estuary is dominated by the large population of intertidal mussels *Mytilus edulis*. The net result is that the overall biomass of macro-invertebrates on the Exe is twice that in Poole Harbour.

This pronounced difference between the two sites is almost certainly due to their different physical characteristics. However, it is also possible that the dominance of mussels on the Exe is not entirely natural, most of the mussel beds having been created by man. A comparison of the bird populations in Poole Harbour and the Exe might reveal whether this major difference between the invertebrate communities is associated with differences in the wintering populations of birds.

Poole Harbour in the past

The intertidal macro-invertebrates of Poole Harbour have been surveyed many times in the past. However, sampling techniques (e.g. sieve mesh sizes) differ between surveys. Thus, the simple comparisons of historical and current data presented here must be treated with some caution. Subject to this caveat, comparison of the three extensive surveys and the four surveys of Holes Bay show that there have been pronounced changes between decades in the abundance of virtually every group of invertebrate. It is impossible to say whether these fluctuations in the invertebrate populations are entirely natural or a response to changed conditions brought about through man's activities in and around the harbour. However, it would seem that change, whatever its cause, is the norm rather than stability.

The future

Pickess and Underhill-Day (2002) demonstrated that of 19 key species of waders and waterfowl, the populations of 10 have increased recently, 5 have been stable and 4 (Shoveler, Oystercatcher, Redshank and Curlew) have declined. There are no explanations for these trends. Although populations of wintering migratory birds in Poole Harbour are determined to a large extent by events and conditions elsewhere, local events and conditions will influence the number of potential recruits that choose to settle and that can survive the winter in good condition. Thus, data local to Poole Harbour will be essential in understanding any continued changes in the population sizes of key bird species.

This survey has established a baseline of macro-invertebrate numerical abundance and biomass. The location of each sampling station has been recorded and each could be revisited to detect changes in prey abundance. Incidental observations indicate that *Mya arenaria* are more abundant in the harbour now than in 2002. The Manila Clam *Tapes philippinarum* is likely to continue its spread within the harbour. A repeat survey is likely to reveal changes. Attributing any such change to human activities will require quantitative data on the important environmental parameters that man's activities might alter, e.g. nutrient inputs, chemical pollution, frequency and intensity of fishing and shellfishing activity and intensity of boat traffic. For a complete analysis, quantitative data on other naturally varying environmental factors, e.g. salinity, sea water temperature and tidal exposure patterns, will also be needed. Understanding the causes of changes to the food supply will be essential in attempting to understand any future changes in the harbour's bird populations.

Acknowledgements

This work was funded by English Nature. We are grateful to Sue Burton who acted as English Nature's Project Manager for this study. We are also extremely grateful to the Royal National Lifeboat Institution and to Tony Stankus in particular for facilitating the survey and increasing the enjoyment levels by allowing us to use the RNLI hovercraft to access many of the sampling locations. We would like to thank Jane Brown, Environmental Co-ordinator at British Petroleum (Wytch Farm) for access to the Institute of Offshore Engineering (1986) report.

References

Arnold, J. B. (1971) *Preliminary Report on the Invertebrate Fauna of Selected Areas in Poole Harbour.* Report to Nature Conservancy Council.

Caldow, R. W. G., McGrorty, S., Durell, S. E. A. le V. dit. and West, A. D. (2003) *Sampling of Benthic Invertebrates in Holes Bay, Poole Harbour.* Report to Wessex Water. Dorchester: Centre for Ecology and Hydrology.

Durell, S. E. A. le V. dit., McGrorty, S., West, A. D., Clarke, R. T., Goss-Custard, J. D. and Stillman, R. A. (2005) A strategy for baseline monitoring of estuary Special Protection Areas. *Biological Conservation,* **121**: 289–301.

Dyrynda, P. (1989) *Marine Biological Survey of the Seabed and Waters of Holes Bay, Poole Harbour, Dorset – 1988.* Report to Dorset County Council. Swansea: University College of Swansea.

Harris, T. (1983a) *An Ecological Survey of Parkstone Bay, Poole, Dorset.* Exeter: University of Exeter.

Harris, T. (1983b) *Holes Bay, Organisms of the Muddy Shore.* Report to Nature Conservancy Council.

Institute of Offshore Engineering (1986) Biological and Chemical Intertidal Survey of Poole Harbour, June 1985. Unpublished report to BP Petroleum Development Ltd. Edinburgh: Heriot-Watt University.

Jensen, A. C., Collins, K. J. and Ovenden, P. J. (1991) *Environmental Monitoring, Poole Harbour 1991. Benthic Survey of Wych Lake and Channel, March 1991.* Report to BP Exploration, Wytch Farm. Southampton: University of Southampton.

McGrorty, S., Rispin, E. and Rose, R. (1987) *Wytch Farm Biological Monitoring: The Invertebrate Fauna of the Intertidal Mudflats at Celeavel Point, Poole Harbour – August, 1987.* Report to BP Petroleum Development Ltd. Wareham: Institute of Terrestrial Ecology.

Pickess, B. P. and Underhill-Day, J. C. (2002) *Important Birds of Poole Harbour*. Wareham: Poole Harbour Study Group.

Thomas, N. S., Caldow, R. W. G, McGrorty, S., Durell, S. E. A. le V. dit., West, A. D. and Stillman, R. A. (2004) *Bird invertebrate prey availability in Poole Harbour.* Wareham: Poole Harbour Study Group.

Warwick, R. M., George, C. L., Pope, N. D. and Rowden, A. A. (1989) *The Prediction of Post Barrage Densities of Shorebirds.* Volume 3: *Invertebrates*. Report to Department of Energy. Plymouth: Plymouth Marine Laboratory (NERC).

8. Sub-tidal Ecology of Poole Harbour – An Overview

Peter Dyrynda

School of Biological Sciences, University of Wales Swansea, Singleton Park, Swansea SA2 8PP
e-mail: P.Dyrynda@swan.ac.uk

Baseline surveys during the 1980s provided the first comprehensive information on the distributions and characteristics of habitats and epibenthic communities within the low water channels of Poole Harbour. Cross- and long-channel variation in habitats and species can be related to equivalent hydraulic gradients involving tidal currents. Dredge, grab and diving assessments collectively produced a species inventory of 68 seaweeds, 159 invertebrates and 32 fish. The best developed epibenthic communities with the greatest biodiversity occur in central areas of the harbour where tidal currents are relatively modest. Peacock Worm *Sabella pavonina* forests are the most significant of the communities found in the harbour, in the context of biodiversity. Human impacts are significant in terms of habitat loss and degradation, e.g. land reclamation, navigational dredging and pollution of various kinds. However, the huge presence and continued arrival of non-native species can be viewed as the most serious threat to biodiversity.

Introduction

Poole Harbour is a 3600 ha, near-land-locked tidal basin located on the central southern coast of England, featuring extensive shores and shoals of sand and mud, dissected and drained by a 35 km network of narrow tidal channels (Figure 1). Although essentially estuarine in character, the harbour also has lagoonal characteristics (Barnes, 1989).

Cotton (1914) considered the sub-tidal channels to be the most interesting feature of the harbour regarding seaweeds, and Waddington (1914) provided sub-tidal invertebrate records, making special mention of the Peacock Worm *Sabella pavonina* "that is to be found in vast numbers" in certain parts of the harbour. Collins and Dixon (1979) undertook the first detailed sub-tidal survey work, using dredging and diving to assess the distribution of oysters and some other conspicuous epibenthic (seabed surface-dwelling) species within an area destined for port expansion.

The need for comprehensive survey data on the harbour as a whole became pressing during the late 1970s and early 1980s as evidence of poor water quality and environmental deterioration coincided with rapid expansion of the Poole conurbation,

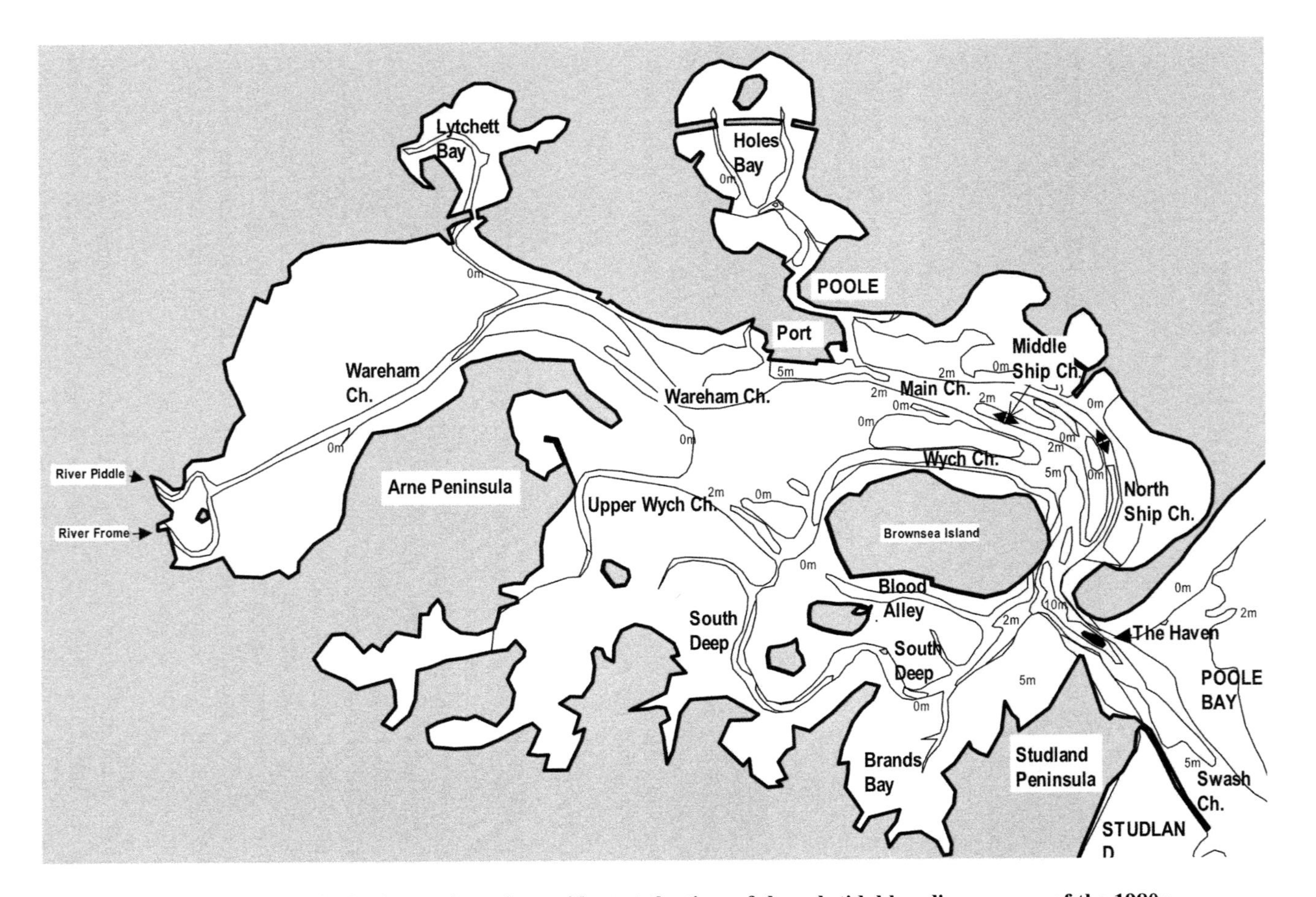

Figure 1 Poole Harbour showing the bathymetric regime evident at the time of the sub-tidal baseline surveys of the 1980s.

further land reclamation for port expansion, marinas, roads and other developments, plans for exploitation of a major oilfield discovered beneath the southern sector, and diversification and intensification of the shellfish culture industry.

In 1982, the sub-tidal channels within Holes Bay were surveyed systematically for the first time (Dyrynda, 1983), followed by those in the southern and northern sectors of the main harbour basin in 1984 and 1985, respectively (Dyrynda, 1985, 1987a, b). These surveys provided a baseline of information on sub-tidal habitats and communities. They revealed the existence of complex distributional patterns of substrates and epibenthos within the tidal channels, and provided indications of the environmental parameters that govern them.

More localized sub-tidal studies undertaken since have included EIA-related work in the South Deep, the port area and Holes Bay, and unpublished studies of the ecology of docks, marinas and farmed shellfish grounds (Dyrynda, 1988, 1989, 1994a, b and others).

In 2003, a new dredging survey was undertaken, covering most areas of the channel system, and providing the first opportunity to gauge ecological stability and change since the baseline surveys two decades earlier (Dyrynda, in prep.).

Howard and Moore (1989) reviewed existing knowledge of the harbour environment in the context of marine conservation, and Langston *et al.* (2003) have provided an up-to-date and detailed environment review, with a particular emphasis on environmental quality.

This chapter provides a synthesis and overview of the habitats and biological communities identified within the sub-tidal zone of Poole Harbour over recent decades, together with a brief analysis of the natural and human factors that influence them.

Methods

The 1982 Holes Bay sub-tidal survey involved channel-centre grab sampling to investigate sediments and infauna, long-channel dredge hauls to appraise larger epibenthic species, and diving to examine fouling assemblages on dock walls (Dyrynda, 1983). The 1984 and 1985 surveys of the main harbour basin involved dredge hauls at 130 locations and grab sampling at a proportion (Dyrynda, 1985, 1987a, b) (Figure 2).

Exploratory dredging at different points across selected channel profiles showed that bottom substrates and key epibenthic species are in many places localized in distribution. To examine these spatial patterns more systematically, 49 cross-channel dive transects were surveyed at points throughout the low water channel network (excluding Holes and Lytchett Bays) (Figure 2). For each transect, a diver swam the length of a survey line recording at 1 m intervals: (i) seabed depth; (ii) substrate and bedform; (iii) key

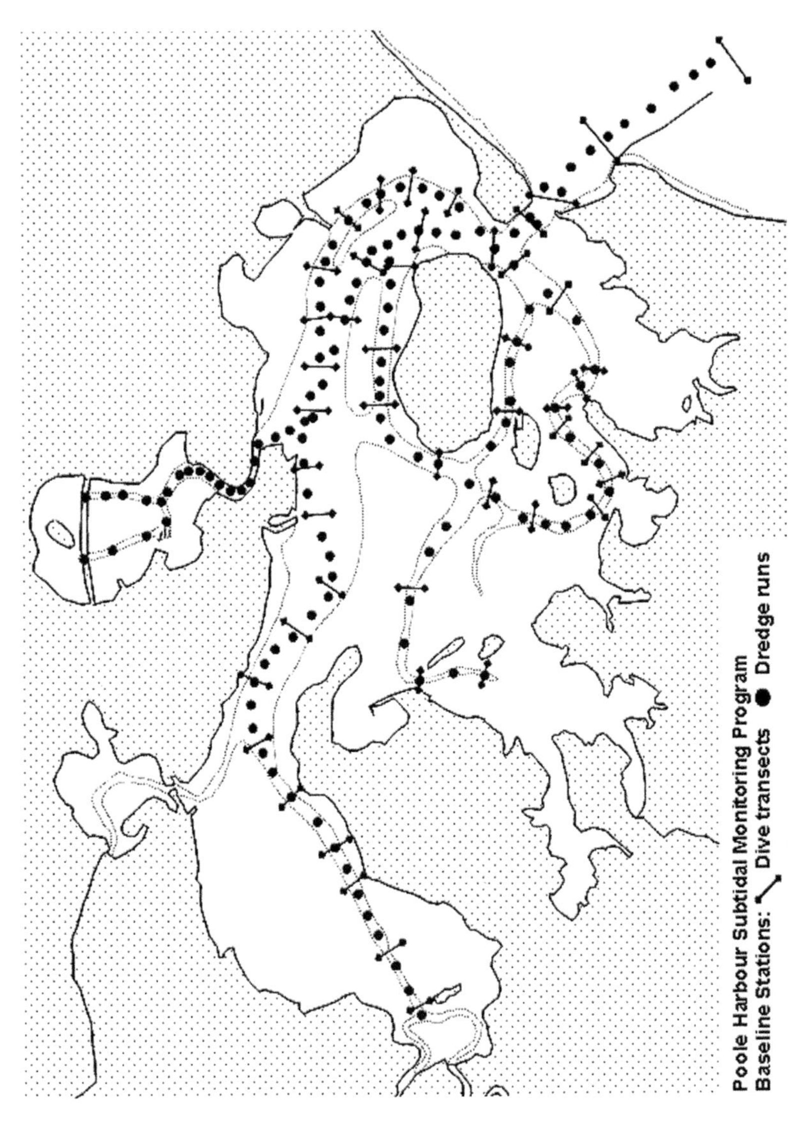

Figure 2 Dredge and dive survey stations assessed during the 1980s baseline surveys.

epibenthic seaweeds and invertebrates; (iv) evidence of human influence. The transects collectively covered more than 5000 m of channel bottom, ranging from the 15.6 m deep harbour entrance (spanned in entirety) to shallow areas less than 1 m deep within upper reaches. All surveys were undertaken during the summer months.

The 2003 dredging programme once again involved long-channel hauls undertaken at channel-centre throughout the harbour. Retrieved samples were sorted into species, an inventory was produced and some of the more abundant species were weighed.

Physical characteristics of Poole Harbour

The main harbour basin, which is broad in comparison with many estuaries contains five islands. Fresh water enters Poole Harbour through several small rivers and streams, the largest being the River Frome. All tidal exchange takes place through a single, narrow tidal inlet to the east flanked by a pair of sandy peninsulas (Figure 1). There are also two subsidiary basins to the north: Holes Bay and Lytchett Bay, which have become increasingly isolated as a result of land reclamation.

The tidal range is small (maximum ~ 2 m) and the cycle asymmetrical, with a double high tide producing prolonged high-tidal stands, followed by a rapid ebb. The salinity regime across much of the harbour is in the range 20–30‰ and quite stable both through the tidal cycle and seasonally, although the Wareham Channel and adjacent shores experience a lower and more variable regime (10–25 ‰).

The sub-tidal channels are generally shallow (mostly <5 m), although they are deepened locally as a result of enhanced tidal scour or navigational dredging (Figure 1). Tidal currents are generally in the 0.5–1 m s^{-2} range, although greater speeds are attained where flow is constricted, notably within the harbour entrance (maximum ~ 2.3 m s^{-2}).

Physical environment within sub-tidal channels – configuration and bathymetry

The main sub-tidal channel lineage extends from Poole Bar up the Swash Channel, Brownsea Roads, the Middle Ship Channel and the Wareham Channel to the mouth of the River Frome (Figure 1). Important subsidiary lineages include the South Deep and Blood Alley, the Wych Channel, the North Ship Channel and those draining Holes and Lytchett Bays (Figure 1).

The Admiralty Charts for Poole Harbour, bathymetric data from Poole Harbour Commissioners, and dive-transect results from the 1980s' baseline surveys all illustrate typical downstream increases in depth and cross-sectional area within each channel lineage, indicative of increasing tidal velocity, volumes and scour (Figure 1). Linear sections of undredged channel usually feature symmetrical profiles with gently sloping flanks. Localized, steep-flanked scour holes reflect increased erosion where tidal flow is concentrated, constricted or deflected, most often within channel bends where the cross-channel profile is typically skewed towards the outer, eroding flank (Figure 1).

Sections of channel subjected to regular navigational dredging are not only 'over-deepened' but also feature unnaturally geometric profiles. During the early 1980s, capital dredging was undertaken alongside land reclamation in the outermost Wareham Channel, to provide a deepwater turning basin for ferries and other vessels. Prior to 1987, the then Main Channel (now the North Ship Channel) was dredged periodically along much of its length. In 1987, capital dredging took place on Poole Bar and along the length of the former Middle or Diver Channel which has since replaced the North Ship Channel as the main navigation route.

Bottom substrates and bedforms

The dive transect baseline surveys of the 1980s revealed the existence of a broad spectrum of substrates across the sub-tidal channel network. In some areas, a succession of different substrates occurred across a single channel transect, whereas in others substrate diversity was low. The dive results also showed that the highest energy substrata typically occur at channel-centre within linear sections of channel, and towards the outer, eroding flanks within channel bends (Figure 2). For example, within the narrowest section of the harbour mouth, medium sand occurs along the channel peripheries, grading down-slope through coarse sand and gravel, to an outcrop of hard clay and adjacent accumulations of rough cobbles and oyster shells within the deepest section. A little further upstream, the eroding eastern flank of Brownsea Island features a steep slope of hard clay, and outcrops of hard 'ironstone'.

The outermost bend within the South Deep featured fine to medium sand upon the inner, depositing flank, A steep slope of hard clay formed the outer, eroding flank. The channel floor was dominated by a Peacock Worm forest.

Data combined for all dive transects show a general upstream-downstream shift from low to high energy substrata, in all channel lineages (Figure 3). Soft mud plains dominate upstream-most sections, whereas coarse sediments (coarse sand, gravel, cobbles), exposures of hard clay and locally, hard bedrock prevail in downstream-most areas. Trains of mobile quartz sand and gravel waves orientated perpendicular to the axis of tidal flow were observed within the lowest sections of the Wych and Middle Channels. Waves observed locally within the middle reaches of the Wareham Channel were composed of mollusc shell gravel.

Fine to medium sands dominate intermediate sections of every channel lineage, often augmented by veneers or scatterings of the shells of smaller molluscs. The species involved vary from area to area. Slipper Limpet and cockle shells are most common within more downstream areas, and shells of the Gaper *Mya arenaria* prevail within the upper reaches of the Wareham Channel.

The combined dive transect data also revealed a south-west to north-east gradient across the harbour of increasingly coarse sediments, that may reflect increased exposure to wave action (with increasing 'wind fetch') (Figure 3).

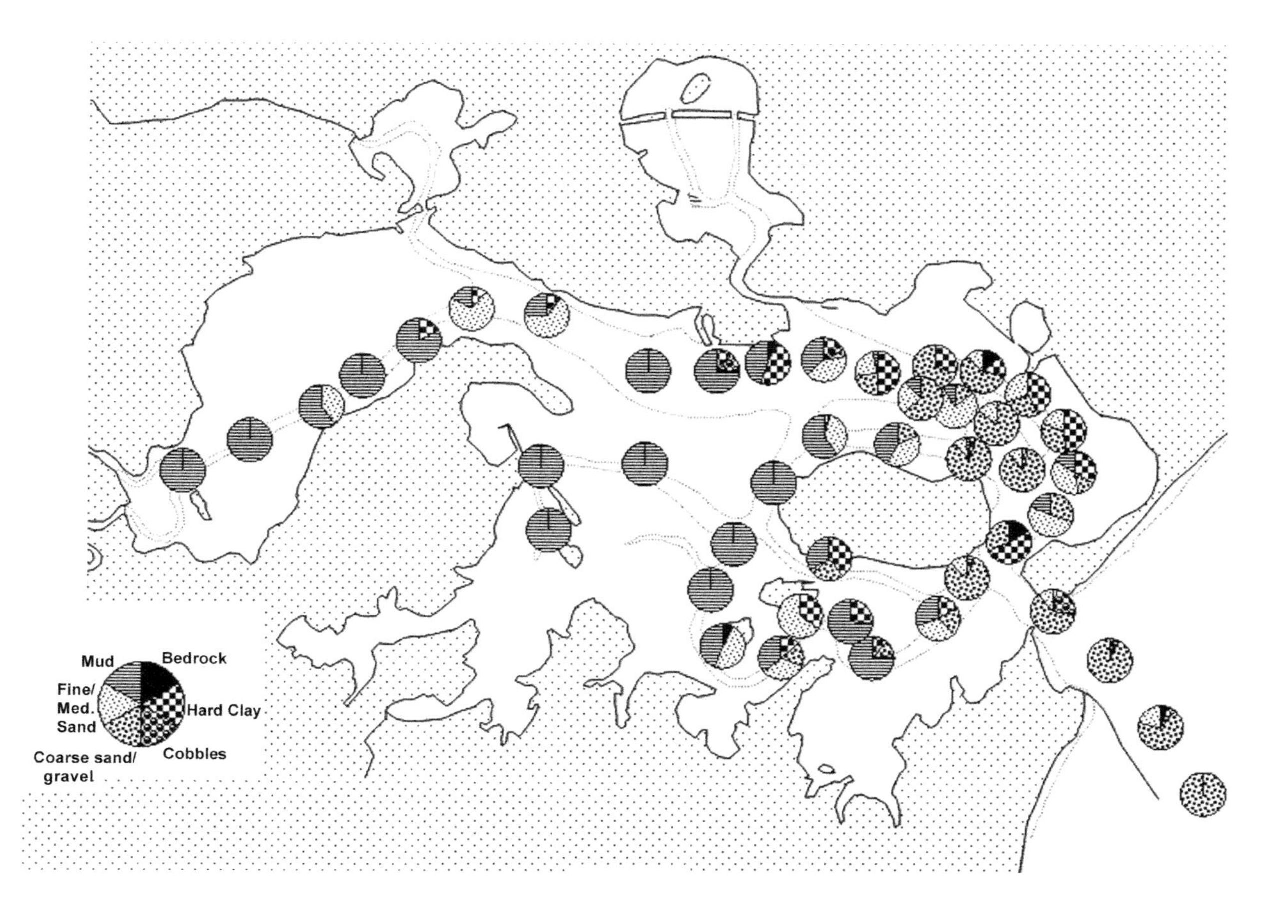

Figure 3 Harbour-wide distribution derived from dive transect data: bottom substrates.

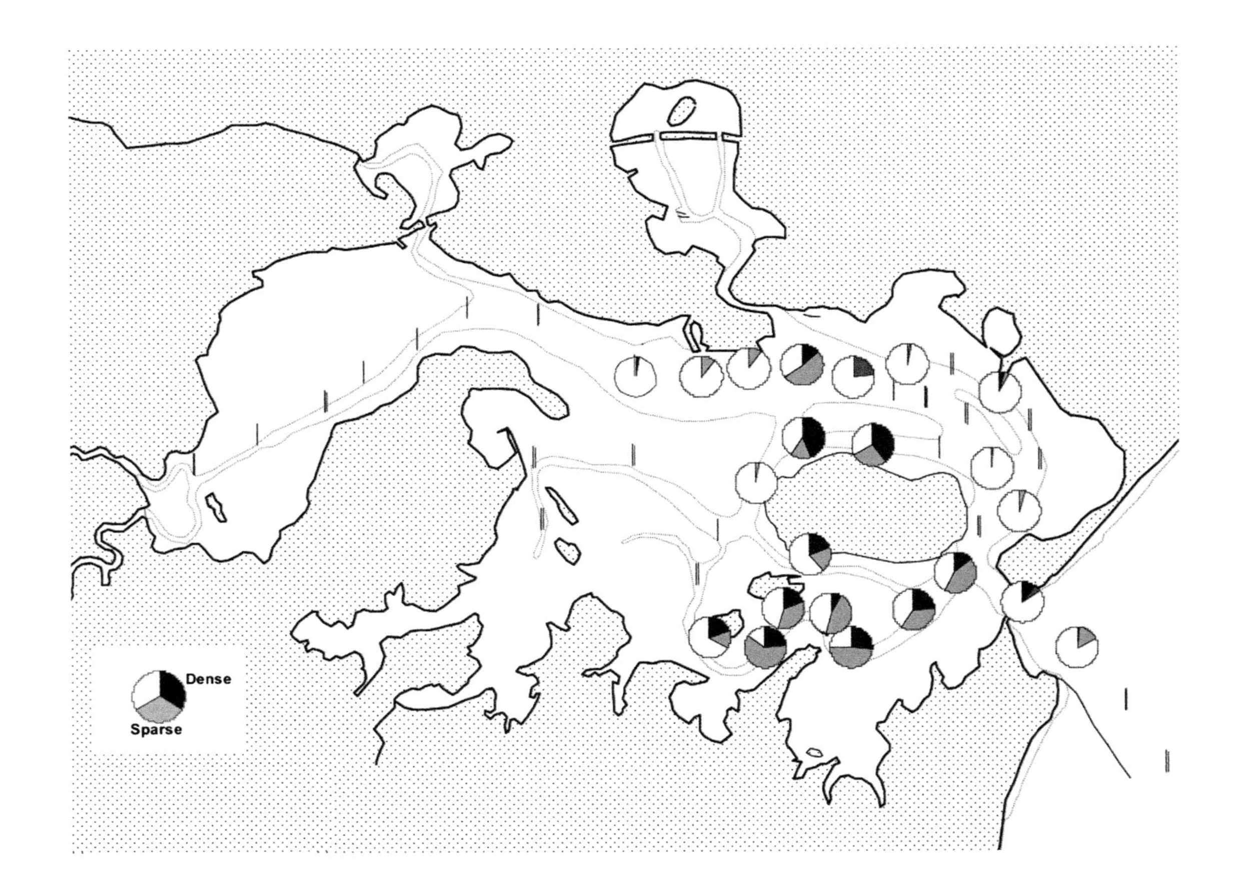

Figure 4 **Harbour-wide distribution derived from dive transect data *Sabella pavonina* forests.**

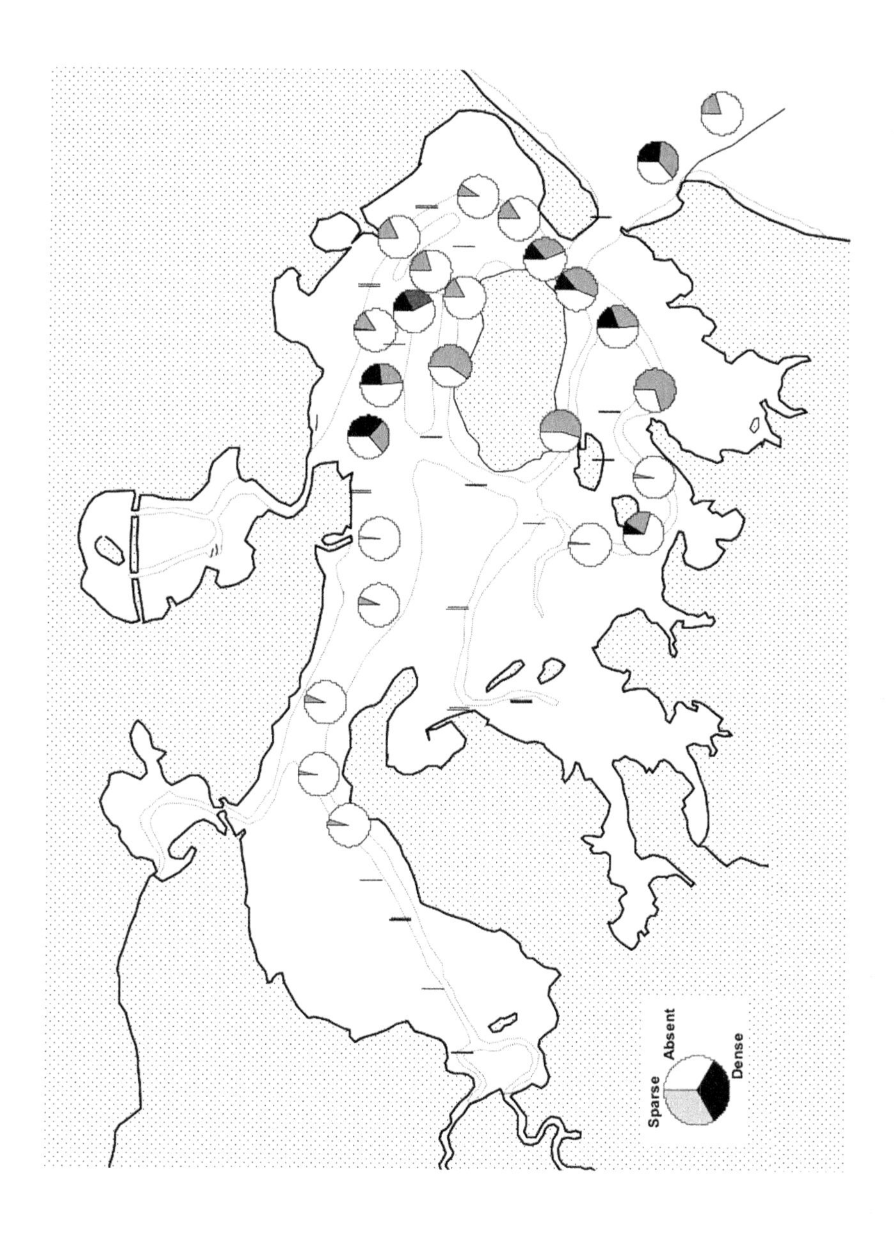

Figure 5 Harbour-wide distribution derived from dive transect data: *Crepidula fornicata.*

Epibenthic communities

Baseline survey results show that the potential for the development of epibenthic communities is influenced by the substrate, bedform and sediment transport induced by tidal currents, in addition to the salinity regime.

In upstream areas, the softness of the sediment and high rates of fine particle accretion are limiting factors. Within current-swept downstream areas, cobbles, stones and oyster shells are liable to be mobilized, and the bed-load transport of sand and gravel can exert severe abrasion pressure. The muds and fine to medium sands characteristic of intermediate tidal energy sections of the channel system proved most favourable for the development of epibenthic communities.

Upstream channel sections

Upstream sections of channel floored by soft mud were found to be largely or wholly devoid of epibenthic macrobiota (seaweeds or invertebrates). Numbers of Common Eel *Anguilla anguilla* burrows were observed within some areas of the channel system.

Mid-stream channel sections

Turbid water assemblages dominated by silt-tolerant invertebrates

Characteristic assemblages of silt-tolerant species were identified towards the top of the zone of epibenthos domination within every channel lineage. Tracts of mud and mud/sand mixtures supported well-separated 'islands' of epibenthic invertebrates including the introduced Korean Sea Squirt *Styela clava*, large growths of the sponges *Cliona celata*, *Haliclona oculata* and *Suberites massa*, and the bryozoan *Anguinella palmata* (Figure 6).

The first European record for *S. clava* was from Plymouth in 1953. It has since spread to many north-east European coasts, as a fouling organism and in association with translocated shellfish (Eno *et al.*, 1997).

Peacock Worm Sabella pavonina forests

The Peacock Worm *Sabella pavonina* was found to form unusually dense and extensive forests within the channels of Poole Harbour (Figure 4). These prevail downstream of the 'turbid water' assemblages. Particularly well-developed forests were found in the South Deep and the Wych Channel. They occur on channel flanks where tidal currents are relatively strong, and on the central channel bottom where currents are more modest.

An adult tube may attain a length of 40 cm, at least half being embedded in sediment. Diving observations showed that fine sediment often settles within these forests, in places forming mudbanks 0.5 m or more in thickness.

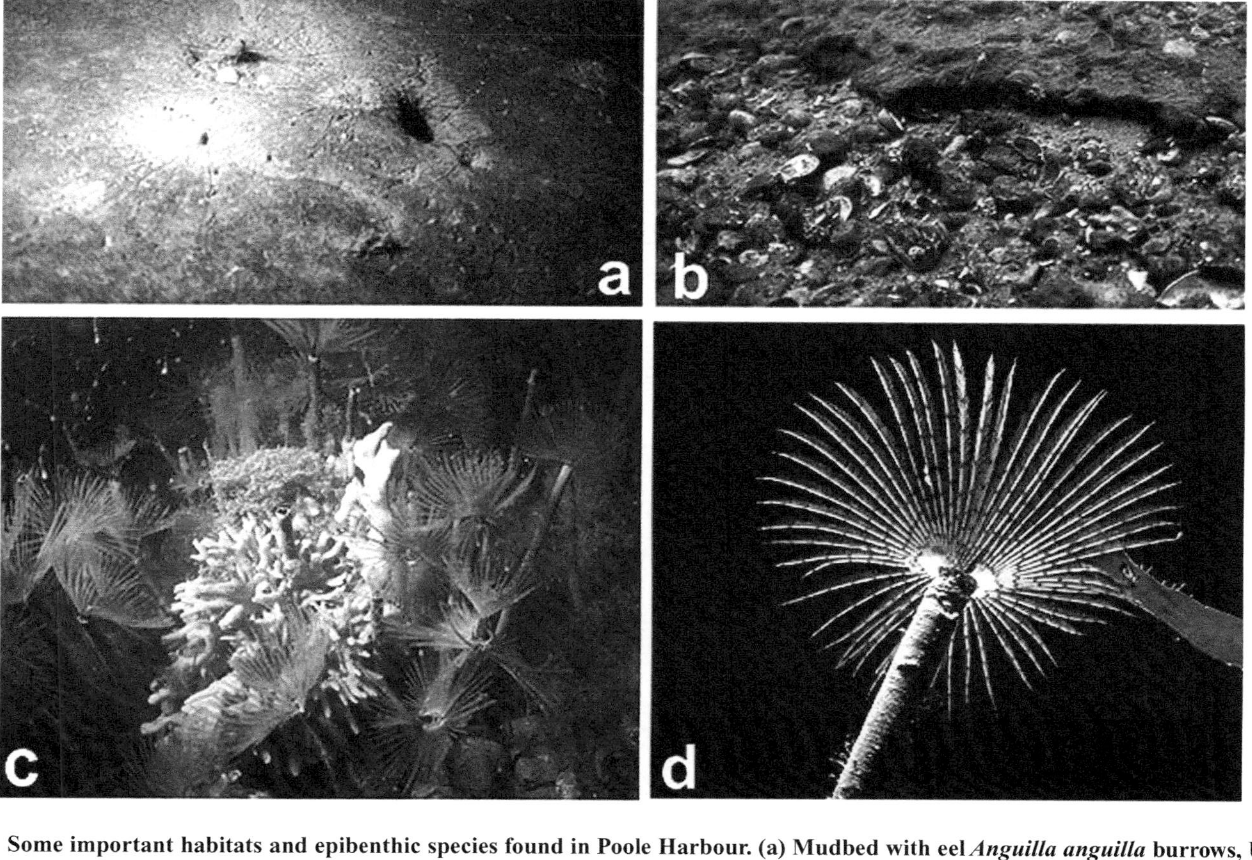

Figure 6 Some important habitats and epibenthic species found in Poole Harbour. (a) Mudbed with eel *Anguilla anguilla* burrows, Upper Wych Channel. (b) Tide scoured hard clay and cobbles, Haven Channel. (c) Peacock Worm *Sabella pavonina*. forest with the sponge *Halichondria bowerbanki*. (d) Tentacle crown of *Sabella*. (Figure 6 continued overleaf)

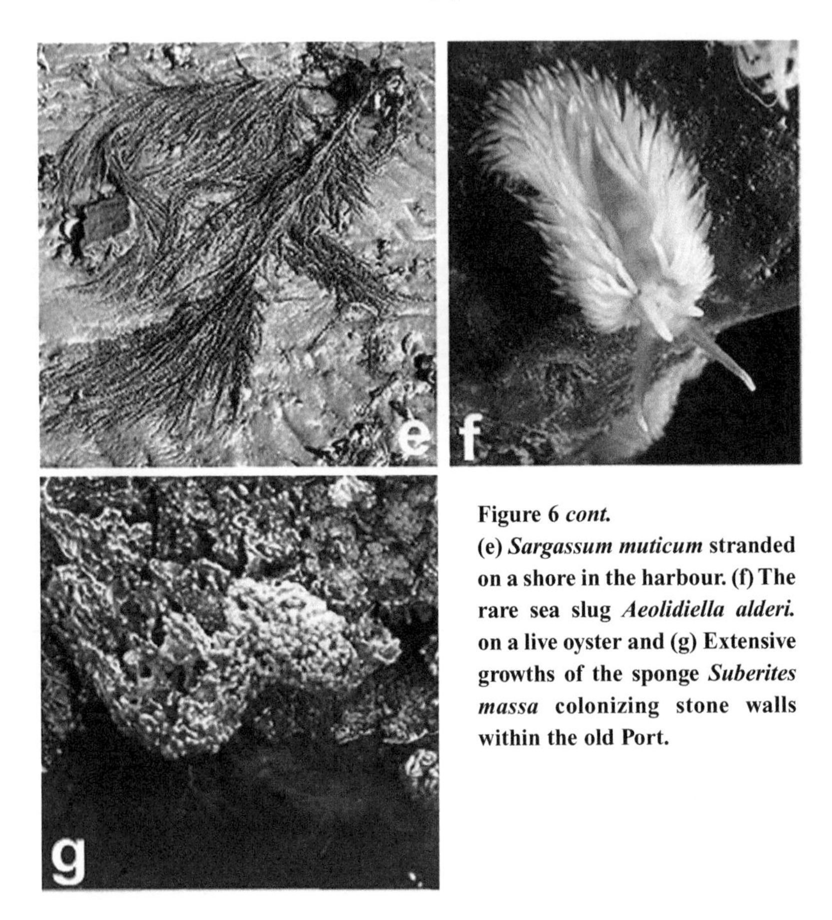

Figure 6 *cont.*
(e) *Sargassum muticum* stranded on a shore in the harbour. (f) The rare sea slug *Aeolidiella alderi.* on a live oyster and (g) Extensive growths of the sponge *Suberites massa* colonizing stone walls within the old Port.

Emergent parts of worm tubes are often colonized by red seaweeds, the sponge *Halichondria bowerbanki*, or clusters of the sea squirt *Ascidiella aspersa*. A multitude of subsidiary seaweeds, invertebrates and fish are also typically present within the *Sabella* forests.

Slipper Limpet Crepidula fornicata beds

The baseline dive transect and dredging baseline surveys identified the widespread occurrence of beds of *Crepidula fornicata* clusters within the low water channel system (Figure 5). Pioneer individuals can colonize various small hard substrates, but subsequent recruits often settle upon established individuals, so forming clusters. The maximum number of limpets observed within a Poole cluster is 27.

High-density limpet beds were found to be widespread towards the downstream end of the intermediate tidal energy zone. The dredge surveys also identified smaller beds within more upstream sections of all channel lineages, where tidal currents and scour are enhanced locally. Limpets were recorded upon a variety of substrata including firm mud, fine to medium sand and hard clay, on level bottoms or within scour holes, but not within the harbour mouth where currents are exceptionally strong.

The Slipper Limpet is a native of the north-east coast of North America. It is believed to have been introduced to UK waters about 1890 on ship hulls and translocated American Oysters *Crassostrea virginica* (Eno *et al.*, 1997).

Sargassum muticum and other seaweeds

Baseline surveys identified large copses of *Sargassum* within the outer harbour, along the margins of the lower South Deep, Wych and Middle Channels, and across the channel bottom within the much shallower Blood Alley. Plants typically grow attached to small hard substrates such as cobbles or the shells of dead or live molluscs.

Field observatons have shown that *Sargassum* plants are routinely transported around the channel system, and in and out of the harbour by tidal currents, particularly during summer months when fronds can attain lengths of several metres. Many mobilized plants drag their anchoring substrates with them, whereas others drift after becoming detached.

Sargassum muticum was first recorded in the UK in the Solent in 1973 (Eno *et al.*, 1997). It was first observed in Poole Harbour in 1979 (personal observation). Other seaweeds encountered frequently within the clearer waters of the outer harbour (in the 1980s and 2003) included *Polysiphonia elongata*, *Gracilaria* spp. and *Dictyota dichotoma* (notably within the lower Wych Channel and Blood Alley). The 2003 dredge survey also identified large quantities of a species of sea lettuce *Ulva rigida* across much of the harbour. During the 1980s, it was only found in quantity in the outermost South Deep. Three previously unrecorded non-native species were also discovered. The red seaweed *Aghardiella subulata* was found in quantity in the South Deep, Blood Alley, Wych Channel and the outer Wareham Channel. It is thought to have arrived in the Solent area prior to 1973 (Eno *et al.*, 1997). *Gracilaria doryhora* was found in the Upper South Deep. This species was first encountered in the UK in 1969 (Eno *et al.*, 1997).

Mature specimens of Wakame *Undaria pinnitifida* were recorded at two locations within the lower Wych Channel in 2003. This large, invasive brown seaweed is of north-east Pacific origin. It was first found in the Poole Harbour area in March 2004 as drift specimens stranded in Shell Bay, and growing attached to floating docks within one harbour marina (personal observation). A substantial build-up of this species seems likely. The first UK record was from the Hamble estuary in 1994, and *Undaria* has since expanded its range and abundance considerably since that time (Eno *et al.*, 1997; Fletcher and Farrell, 1999).

Downstream channel sections – tide-swept communities

Substrata within the most tide-swept sections of the channel system in and around the harbour entrance support an impoverished community characterized by scour-tolerant forms. Within one area, scour and abrasion are extremely intense, and boulders embedded in gravel were polished smooth, and barren. Elsewhere, cobbles and oyster shells supported low-drag, abrasion- and impact-resistant species; typically heavily calcified invertebrates such as the tube worm *Pomatoceros lamarcki*, the barnacle *Balanus crenatus* and the encrusting bryozoan *Cryptosula pallasiana*, and characteristic higher-drag species such as the hydroid *Hydrallmania falcata* and the bryozoans *Walkeria uvae*, *Alcyonidium diaphanum* and *Flustra foliacea*.

Localized accumulations of rough cobbles were found to be sufficiently stable to support larger, higher drag species, e.g. *Sargassum muticum* on one flank of Stone Island, and unexpectedly, a sparse cover of Peacock Worms within the deepest section of the harbour mouth.

Hard clay was generally free of attached species, but was in places bored by the North American Piddock *Petricola pholadiformis*. This species is thought to have arrived in the UK from North America before 1890 (Eno *et al.*, 1997).

High energy coarse sand and gravel plains and also wave trains within and beyond the harbour were largely free of epibenthos apart from quantities of drift algae deposited on the bottom during slack tide periods, especially *Sargassum muticum* and *Laminaria saccharina* (Figure 6). Motile species encountered in these areas included the Hermit Crab *Pagurus bernhardus*, the Harbour Crab *Liocarcinus depurator*, the Common Cuttlefish *Sepia officinalis* and the Lesser Sand-eel *Ammodytes tobianus*.

Humanized environments

Dredged shipping channels

Recently dredged areas within the North Ship Channel and outermost Wareham Channel were found to be largely barren, apart from scattered biota rafted in by tidal currents. At one North Channel location, fragments of a Peacock Worm forest were evident, dissected by dredge-excavated trenches.

A dive transect spanning the Main Channel assessed one day after grab dredging, revealed relatively intact tracts of channel bottom with associated epibenthic seaweeds and invertebrates interrupted by 'barren' dredge craters.

The 2003 dredge survey was of interest in the context of recovery after navigational dredging. The survey revealed a substantial presence of Slipper Limpets in the North Ship Channel colonized by the sea squirt *Ascidiella aspersa* (dredging ceased around 1987).

Natural and farmed oyster beds

During the baseline surveys of the 1980s, and also the 2003 re-appraisal, numbers of 'wild oysters' in the sub-tidal channels were found to be low, with sporadic 'ones and twos' being typical of hauls in favourable areas (apart from high densities encountered where stock had been deposited within leased grounds).

Diving observations indicate that intensive shellfish dredging can have a significant impact on channel bottom sediments and biological assemblages. Shellfish dredges typically retain a substantial by-catch of non-target algae and sedentary invertebrates (hence their value for survey work).

Sub-tidal shellfish grounds are usually cleared of 'undergrowth' before laying new stock. This process usually mobilizes fine sediment, exposing previously buried shell material and coarse sediment fractions at the surface.

At the time of the baseline surveys, Solent oysters deposited in the South Deep were found to be colonized by the sea squirt *Dendrodoa grossularia*, which does not occur naturally in the harbour. Monitoring of deposited oysters and their epibionts in 1988–89 showed that *Dendrodoa* and other fully saline species died-off progressively and were replaced by colonizers typical of the harbour. History has shown that oysters can serve as Trojan horses for diseases and non-native species, and short-distance translocations can play an important role in the shuttling of alien species around a region, whereas long-distance translocations (trans- or inter-oceanic) can lead to totally new species introductions (Eno *et al.*, 1997).

Baseline surveys and other studies conducted on leased grounds suggest that farmed native oyster beds can become biodiverse. The Sulphur Sponge *Cliona celata* and the tubeworm *Polydora ciliata* often burrow into the shells of oysters. The sea anemone *Cereus pedunculatus*; the Keel Worm *Pomatoceros triqueter*, barnacles (including the non-native *Elminius modestus*), bryozoans (particularly *Conopeum reticulum* and *Cryptosula pallasiana*) and sea squirts (particularly *Ascidiella aspersa* and the Korean Sea Squirt *Styela clava*) adorn the shells of live and dead oysters, and a variety of small crabs and fish use gaping shells as refuges.

A population of the rare sea slug *Aeolidiella alderi* was identified on an experimental oyster plot west of Brownsea Island. Individuals were seen consuming sea anemones and laying their eggs upon oyster shells (unpublished observation).

Docks and marinas

The port area and small vessel marinas contain a variety of seawalls, jetties, floating docks, buoys and fixed channel markers, which provide good substrates for colonization by seaweeds and invertebrates. The sponge *Suberites massa* occurs in large numbers upon submerged dock walls within the channel leading into Holes Bay (Figure 6). This population is probably the most substantial found in UK waters. Other common organisms include the sponges *Halichondria bowerbanki*, *Haliclona oculata* and *Microciona atrasanguinea*, the sea anemone *Metridium senile*, the barnacle *Elminius modestus* and the Korean Sea Squirt *Styela clava.*

Marina pontoon floats and the unprotected hulls of vessels support assemblages that are significantly different to those found on fixed structures. Typical species include *Sargassum muticum*, the sponge *Sycon ciliatum*, the sea squirts *Botryllus schlosseri* and *Clavelina lepadiformis* and the non-native bryozoan *Tricellaria inopinata*, a Pacific species first recorded in Poole Harbour in 1998, representing only the second North Atlantic record (Dyrynda *et al.,* 2000).

Discussion

The dive transect method devised for the 1980s baseline surveys of the Poole Harbour proved very effective for the rapid assessment of the distributions of substrates, bedforms and epibenthic assemblages in relation to bathymetry. The role of hydraulic forces in governing the distribution and nature of epibenthic assemblages in coastal waters is well recognized (Hiscock, 1983), although the information gathered by this study on small-scale patterns within estuarine channels is relatively novel. The same is true of the identification of miniature equivalents to substrate regimes and bedforms normally associated with tide-swept, offshore shelf sea environments, e.g. sand and gravel waves and ribbons (Stride, 1982).

Biodiversity

Comparisons of the biodiversity (measured as species-richness – the number of species) of harbours and other marine inlets in southern Britain indicate that Poole Harbour is less diverse than certain fully saline inlets in Devon, Cornwall and Pembrokeshire, but more diverse than many others (Howard and Moore, 1989; Davidson *et al.*, 1991). The combined dredge, grab and dive transect baseline survey revealed the presence of 68 seaweed species, 159 invertebrates and 32 fish (Dyrynda, 1987 a, b). The list is significantly different to the inventory compiled by Mallinson *et al.* (1999) during their long-term artificial reef studies in the fully saline waters of Poole Bay.

Peacock worm forests have been identified as 'biodiversity hot spots' within the harbour. *Sabella pavonina* is a widespread and common species in British coastal waters, but densities are generally low. The dense and extensive aggregations in Poole Harbour are exceptional. An abundant food supply in the form of organic detritus descending from the saltmarshes and other intertidal habitats, combined with moderate to strong tidal currents (enhancing the volume of water filtered), and extreme shelter from wave action are perhaps responsible for these forests.

Biological productivity

Poole Harbour is a highly productive environment, with prolific growth of seaweeds and saltmarsh plants generating a favourable food supply for a multitude of deposit and suspension-feeding invertebrates inhabiting shores, shoals and channels, including harvested clams, cockles, mussels and oysters, but also sponges, Peacock Worms and sea squirts. This production also passes through the food web to higher consumers including fish and shore birds of conservation and/or economic significance.

Rare species

Howard and Moore (1989) drawing on species lists produced during previous baseline surveys highlight the sponge *Suberites massa* and the bryozoans *Anguinella palmata*, *Farella repens* and the sea slug *Aeolidiella sanguinea* as being of national significance in terms of rarity.

The likelihood is that the first three species are generally under-recorded by UK surveys. The grey muddy colonies of *Anguinella* are quite large but easily overlooked. *Farella repens* is a miniscule but distinctive species requiring careful microscopic examination. Although the 'brain-like' mound colonies of *Suberites massa* that abound within the Port of Poole are very distinctive (Figure 6), this species can also grow as encrusting patches that bear a superficial resemblance to *Hymeniacidon perleve*, a common shore species. In view of the occurrence of all three species in dock locations, their status should be viewed as 'cryptogenic', i.e. it is not clear as to whether they are native or non-native. As reported by Dyrynda (1991), it is likely that the single sea slug collected in the outer South Deep during the baseline surveys and identified as *Aeolidiella sanginea* (Dyrynda, 1985) is in fact *Aeolidiella alderi*, less uncommon but still viewed as rare in UK waters.

Natural habitat loss

Sub-tidal losses are most evident with the port area, along the sides of the channel leading into Holes Bay, and within the outermost section of the Wareham Channel, and have involved sedimentary seabed areas (mud, sand and shingle).

With reference to gains, harbour walls, breakwaters, pilings and other fixed structures serve as artificial reefs colonized heavily by a diversity of seaweeds and invertebrates. Pontoon floats located within the harbour's many small vessel marinas also support rich populations of seaweeds, invertebrates and fish, the composition being significantly different to those of fixed structures.

The benefits of artificial reefs are well-known from the studies elsewhere, including work undertaken by Southampton University on a constructed reef in Poole Bay (Mallinson *et al.*, 1999 and others). However, one potential disbenefit of docks and marinas as artificial reefs is their obvious role as relay points and incubators for non-native species (see below).

Habitat degradation

Navigational dredging

Navigational dredging represents a radical form of physical disturbance regarding benthic habitats and the benthos in that both are removed totally. Survey data from the harbour suggests that recovery may be slow and the end-point habitats and communities may well be quite different to the natural ones prior to dredging. Dredging is, however, confined to certain channels, leaving others unaffected.

Dredging for shellfish ground clearance and stock harvesting

Oyster dredging represents a much less radical form of disturbance than the aforementioned, although the impact of intensive shellfish dredging can be significant. Intensive dredging favours robust species such as the Slipper Limpet at the expense of relatively sensitive forms such as the Peacock Worm. Intensive dredging is confined to limited areas operated by leaseholders leaving most of the channel system relatively undisturbed.

Chemical disturbances

The poor flushing characteristics of Poole Harbour in general and Holes Bay in particular are well recognized, as is the associated vulnerability regarding pollution (Falconer, 1986; Langston *et al.*, 2003).

Eutrophicating pollutants

Plant nutrients arising from fertilizers, detergents, sewage, silage and other animal wastes have caused major problems in many estuaries and lagoons around the world. Eutrophication can have a deleterious effect on biodiversity. Concerns about excessive growth of the Sea Lettuce *Ulva lactuca* were already being expressed during the 1960s and 1970s, particularly in the context of Holes Bay (Savage, 1971; Portsmouth Polytechnic, 1981; Langston *et al.*, 2003). The more recent, major build-up of *Ulva rigida* across the main harbour basin would suggest further increases in enrichment have occurred. As well as suppressing biodiversity, there is the risk of crises comparable with those seen in Italian lagoons, involving mass mortalities of invertebrates smothered by live or decaying seaweeds or asphyxiated by low dissolved oxygen levels associated with high rates of decay of organic matter. Mass mortalities of mussels in the harbour during the mid-1990s may have been linked to this, although thermal stress is considered a more likely causative factor (Dyrynda and Brown, 1998).

Exceptionally high frequencies of shell disease reported for the Brown Shrimp *Crangon crangon* in Poole Harbour channels may represent another consequence of eutrophication (Dyrynda, 1998).

Toxic metals and organic substances

Chemical pollution has been a recognized problem within Holes Bay since the 1970s. A succession of studies has identified elevated levels of cadmium, mercury and other toxic metals within both sediments and organisms (Boyden, 1975; Langston, 1982; Langston *et al.*, 1987, 2003). During the 1980s, exceptionally high concentrations of tributyl tin (then much used in antifouling paints) were recorded in Holes Bay, and abnormal gross shell thickening in farmed stock of the Pacific Oyster *Crassostrea gigas*, a bioindication of TBT pollution, proved problematical in the south side of the main harbour basin. The latter was alleviated by the 1987 partial ban on the use of TBT antifouling paints on vessel hulls (Dyrynda, 1992). It is known that TBT affects a much wider range of organisms than just oysters, although most are of no economic significance. It is worth noting that the baseline surveys of the 1980s were undertaken at a time when TBT levels in the harbour were very high. The effects of this on the baseline species inventory remain unknown.

Unsustainable exploitation of living natural resources

Natural stocks of harvestable marine life within the confines of Poole Harbour are limited in scale. Flat Oyster populations, now at a fraction of their historical levels, were undoubtedly overfished in the past, although the tradition during the 1800s was to deposit a proportion of stock harvested from open coastal waters within the channels of the harbour (Philpots, 1890). Although harvesting is now regulated, disease outbreaks, the presence of the North American Slipper Limpet and overfishing have all hampered the natural recovery of this species.

Non-native species introductions

The baseline surveys of the 1980s revealed a huge presence of non-native species in the tidal channels of the harbour, and subsequent monitoring shows that new species continue to arrive on a regular basis. These non-native species are causing major changes to the structure of natural communities within the low water channels, with at least partial displacement of 'native' biota. The North American Slipper Limpet *Crepidula fornicata* is the dominant epibenthic species in the harbour channels. It clearly out-competes the native Flat Oyster *Ostrea edulis* and now occupies large areas of the seabed that were probably favoured by the European Flat Oyster in previous times.

The impacts of some of the other arrivals are less easy to assess, and some argue that biodiversity is being augmented rather than depressed. For example, *Sargassum muticum* provides a substratum for a multitude of epiphytic algae and invertebrates.

At least six new arrivals have been documented for the harbour over the past 6 years or so including several during the 2003 dredging survey. Most invaders go unnoticed until a specialist survey discovers them, although the large brown seaweed known as Wakame *Undaria pinnitifida*, now in early stages of consolidation, may become almost as conspicuous as *Sargassum muticum*.

In view of the long maritime history of the port, it is likely that many species were introduced before records began. The old sailing ships were often very heavily fouled, and were ideal carriers for non-native biota. For several centuries Poole conducted intensive trade with Newfoundland and New England. Poole Oyster Company was importing live oysters from North America and southern Europe during the late 1800s (Philpots, 1890) and it is certain that other species would have been introduced with them.

The presence of a large number of non-native species adds a tier of instability and complexity to the system as invaders wax and wane over different time-scales, presenting real challenges for the analysis of stability and change in the harbour environment.

Climate change

Impacts of rapidly rising sea levels and temperatures may be substantial within Poole Harbour over the coming decades and centuries. Rising sea levels may ultimately cause enlargement of the harbour, perhaps resulting in ecologically significant changes to the hydraulic regime. Recent hot summers such as that of 1995 may have provided an indication of some ecological impacts likely to be caused by sustained warming. Exceptional mortalities of farmed mussels and unusual blooms of nuisance algae were documented in 1995 (Dyrynda and Brown, 1998). Higher temperatures are likely to be accompanied by distributional shifts in the north-east Atlantic as species adapted to both colder and warmer conditions shift northwards.

Acknowledgements

The author is indebted to the many individuals from Poole and elsewhere, too numerous to mention, who have helped past projects in various ways, including staff from English Nature (and the Nature Conservancy Council before) in Dorset; Poole Harbour Commissioners, including successive crews of the workboat 'Rough Ryder'; the Poole fishing community, notably David and Ian Davies and Chris Brewer who helped directly with field surveys; Poole Borough Council and Dorset County Council; the Environment Agency; members of the Poole Harbour Study Group, and colleagues at University of Wales Swansea, particularly Elisabeth Dyrynda who helped greatly with a number of studies.

References

Barnes, R. S. K. (1989) The coastal lagoons of Britain: an overview and conservation appraisal. *Biological Conservation*, **49**: 295–313.

Boyden, C. R. (1975) Distribution of some trace metals in Poole Harbour, Dorset. *Marine Pollution Bulletin*, **6**: 180–187.

Collins, K. J. and Dixon, I. M. (1979) *A Survey of the Oyster Population off the Hamworthy Shore, Poole Harbour, October 1979*. A report by the University of Southampton to Poole Harbour Commissioners.

Cotton, A. D. (1914) Marine algae. pp. 190–198. In: *A Natural History of Bournemouth and District*. Morris, D. (ed.). Bournemouth: The Natural History Society.

Davidson, N. C., d'Laffoley, D., Doody, J. P., Way, L. S., Gordon, J., Key, R., Pienkowski, M. W., Mitchell, R. and Duff, K. L. (1991) *Nature Conservation and Estuaries in Britain*. Peterborough: Nature Conservancy Council.

Dyrynda, E. A. (1992) Incidence of abnormal shell thickening in the Pacific oyster *Crassostrea gigas* in Poole Harbour (UK), subsequent to the 1987 TBT restrictions. *Marine Pollution Bulletin,* **24**: 156–163.

Dyrynda, E. A. (1998) Shell disease in the common shrimp *Crangon crangon*: variations within an enclosed estuarine system. *Marine Biology*, **132**: 445–452.

Dyrynda, P. E. J. (1983) *Investigation of the Subtidal Ecology of Holes Bay, Poole Harbour*. Report by Swansea University for the Nature Conservancy Council.

Dyrynda, P. E. J. (1985) *Poole Harbour Subtidal Survey – Southern Sector – 1984*. Report by Swansea University for the Nature Conservancy Council.

Dyrynda, P. E. J. (1987a) *Poole Harbour Subtidal Survey – Dive Transect Survey – Northern Sector – 1985*. Report by the University of Wales Swansea for the Nature Conservancy Council.

Dyrynda, P. E. J. (1987b) *Poole Harbour Subtidal Survey Baseline Assessment*. Report by the University of Wales Swansea for the Nature Conservancy Council.

Dyrynda, P. E. J. (1988) *Subtidal Monitoring, South Deep, Poole Harbour*. Report by Swansea University for B.P. Petroleum Development.

Dyrynda, P. E. J. (1989) *Marine Biological Survey of the Bed and Waters of Holes Bay – 1988*. Report by the School of Biological Sciences, Swansea University to Dorset County Council.

Dyrynda, P. E. J. (1991) *Benthic Habitats and Species of Conservation Interest Within Poole Harbour (Southern England)*. Report by University of Wales Swansea for the Nature Conservancy Council.

Dyrynda, P. E. J. (1994a) Hydrodynamic gradients and bryozoan distributions within an estuarine basin (Poole Harbour, UK). pp. 57–64. In: *Biology and Palaeobiology of Bryozoans*. Hayward, P. J., Ryland, J. S. and Taylor, P. D. (eds). Fredensberg: Olsen & Olsen.

Dyrynda, P. E. J. (1994b) *Poole Town Quay Boat Haven Impact Assessment: Marine Biological Survey*. Report to Terence O'Rourke plc and Poole Harbour Commissioners.

Dyrynda, P. E. J. and Brown, F. E. (1998) *Factors Affecting Condition and Mortality of Farmed Mussels in Poole Harbour: 1995–1997*. Report by University of Wales, Swansea for the Southern Sea Fisheries District Committee.

Dyrynda, P. E. J., Fairall, V. R., Occhipinti Ambrogi, A. and d'Hondt, J-L. (2000) The distribution, origins and taxonomy of *Tricellaria inopinata* d'Hondt & Occhipinti Ambrogi, 1985, an invasive bryozoan new to the Atlantic. *Journal of Natural History,* **34**: 1993–2006.

Eno, N. C., Clark, R. A. and Sanderson, W. G. (eds) (1997) *Non-native Species in British Waters: A Review and Directory*. Peterborough: Joint Nature Conservancy Council.

Falconer, R. A. (1986) Water quality simulation in a natural harbour. *Journal of Waterway, Port, Coastal and Ocean Engineering*, **112**: 15–34.

Fletcher, R. L. and Farrell, P. (1999) Introduced brown algae in the North Atlantic, with particular respect to *Undaria pinnatifida* (Harvey) Suringar. *Helgolander Meeresuntersuchungen Helgolander Meeresunters*, **52**: 259–275.

Hiscock, K. (1983) Water movement. pp 58–96. In: *Sublittoral Ecology*. Earll, R. and Erwin, D. G. (eds). Oxford: Clarendon Press.

Howard, S. and Moore, J. (1989) *Surveys of Harbours, Rias and Estuaries in Southern Britain: Poole Harbour*. A Report by the Field Studies Council Oil Pollution Research Unit. Nature Conservancy Council CSD Report No. 896.

Langston, W. J. (1982) The distribution of mercury in British estuarine sediments and its availability to deposit feeding bivalves. *Journal of the Marine Biological Association of the UK*: 667–684.

Langston, W. J., Burt, G. R. and Zhou, M. (1987) Tin and organotin in water, sediments and benthic organisms of Poole Harbour. *Marine Pollution Bulletin*, **18**: 634–639.

Langston, W. J., Chesman, B. S., Burt, G. R., Hawkins, S. J., Readman, J. and Worsfold, P. (2003) Site characterisation of the South West, European Marine Sites. Poole Harbour SPA. *Occasional Publication Marine Biological Association*, No. 12.

Mallinson, J. J., Collins, K. J. and Jensen, A. C. (1999) Species recorded on artificial and natural reefs, Poole Bay, 1989–1996. *Proceedings of the Dorset Natural History & Archaeological Society*, **121**: 113–122.

Philpots, J. R. (1890) *Oysters, and All About Them*. Leicester: John Richardson.

Portsmouth Polytechnic (1981) *Ulva and Enteromorpha in Poole Harbour and Holes Bay*. Portsmouth: Department of Biological Sciences, Portsmouth Polytechnic.

Savage, P. D. V. (1971) *The Seaweed Problem at Poole Power Station*. Central Electricity Research Laboratories.

Stride, A. H. (1982) *Offshore Tidal Sands*. London: Chapman & Hall.

Waddington, H. J. (1914) Marine fauna. Invertebrata. pp. 213–230. In: *A Natural History of Bournemouth and District*. Morris, D. (ed.). Bournemouth: The Natural History Society.

The Ecology of Poole Harbour
John Humphreys and Vincent May (editors)

9. Zooplankton Distribution in Poole Harbour

Paola Barbuto[1], Eunice Pinn[1] and Antony Jensen[2]

[1]School of Conservation Sciences, Bournemouth University, Poole, Dorset BH12 5BB

[2]School of Earth and Ocean Sciences, Southampton Oceanography Centre, University of Southampton, Hampshire SO14 3ZH

The distribution and abundance patterns of the zooplanktonic communities in the estuarine environment of Poole Harbour, and their link with physico-chemical parameters, are reported here for the first time. Over the summer of 1999, the harbour maintained a rich zooplankton standing stock, with peak production during August and September. Copepods and the larvae of gastropods, polychaetes, bivalves and decapods contributed a major share of the zooplankton community. Zooplankton spatial distributions were thought to be partly influenced by the hydrographic structure of the area and by the tidal currents, which contribute to a certain degree to species retention. Zooplankton abundance followed a similar pattern to chlorophyll 'a' concentration, but with a time lag of approximately one month. While spatial distribution patterns may be related to the hydrographic features of the harbour, temporal occurrence of the different zooplankton groups appeared to be related to the autochthonous production of the harbour.

Introduction

As with most coastal ecosystems, estuaries and lagoons are often productive environments supporting fisheries of commercial and recreational importance (e.g. Jouffre *et al.*, 1991; Wilson, 2002). Zooplankton abundance and biomass are both important determinants in assessing the rates of production, the nature of food webs and trophic transformations (Collins and Williams, 1982). Therefore, quantifying and understanding the processes controlling abundance and production of zooplankton within those ecosystems is of great conservation interest.

Planktonic communities in estuaries are typically rich and diverse, and can frequently be enriched by benthic algae resuspended from the seabed (Raymont, 1963). Although phytoplankton is usually abundant, the same does not always appear to be true for zooplankton, whose distribution and abundance can be irregular and are poorly understood (Barnes, 1974).

The inshore zooplankton community is typically characterized by a few holoplanktonic species (i.e. organisms that spend their entire life in the planktonic community) and a considerable variety of meroplanktonic animals (i.e. organisms that spend only part of their life cycle in the plankton, usually as eggs or larvae). The main results of research on zooplankton in coastal waters were reviewed by Colombo *et al.* (1984), who emphasized the role of the biological communities in the coastal ecosystem organization. Ambrogi *et al.* (1987) suggested that the organization and seasonal occurrence of zooplankton in a lagoon depended on autochthonous production, while Collins and Williams (1981) suggested that the spatial distribution of the holoplanktonic community of an estuarine area is mainly dependent on the hydrology (e.g. salinity) of the system.

The ecology of Poole Harbour zooplankton is largely unknown. Therefore, the current study presents baseline information to evaluate production and community structure of zooplankton in the harbour.

Methods

In this study, only the mesoplankton community (animals ranging from 200 μm to 2 mm in size) was considered. This includes both species from the holo- and the meroplankton communities. Sampling was carried out at monthly intervals from May to September in 1999, over a designated horizontal transect of seven systematically distributed stations (Figure 1) at two different depths (1 m and 5 m). Fixed-depth sampling was performed at similar tidal states (high spring tide) to assess spatial distribution patterns of zooplankton within the harbour, whilst changes in population density and composition were monitored through horizontal and vertical tows. The relationship between the zooplankton groups and environmental parameters was assessed using data resulting from the direct measurement of the physico-chemical variables including temperature, salinity, chlorophyll 'a' and phaeo-pigments (Barbuto, 1999).

Results and discussion

The zooplankton community within Poole Harbour showed clear variations in spatial and temporal distribution patterns. Thirty-three groups of zooplankton were identified with clear numerical dominance of holoplanktonic species (Table 1), particularly calanoid copepods (*Acartia* spp.), which is typical of most neritic and estuarine environments (Raymont, 1963). These species were restricted to Wareham Channel, which suggests that there was a certain degree of retention of the population within this part of the harbour.

With regard to the meroplanktonic groups, cirripede nauplii and the larvae of gastropods, bivalves, polychaetes and decapods were the most abundant groups, particularly in the Upper Wych Channel (Table 1). Their spatial distribution appeared to be partly influenced by the location of the adult population (Burkill, 1983), and by the hydrographic features of the system. The Upper Wych Channel (sampling stations 3 and 4) (Figure 1) borders molluscan aquaculture beds (Manila Clam, Palourde, oyster, cockle and mussel) in this part of the harbour.

The planktonic assemblages maintained substantially the same composition from May to September (Figure 2). The communities belonging to defined areas of the harbour appeared

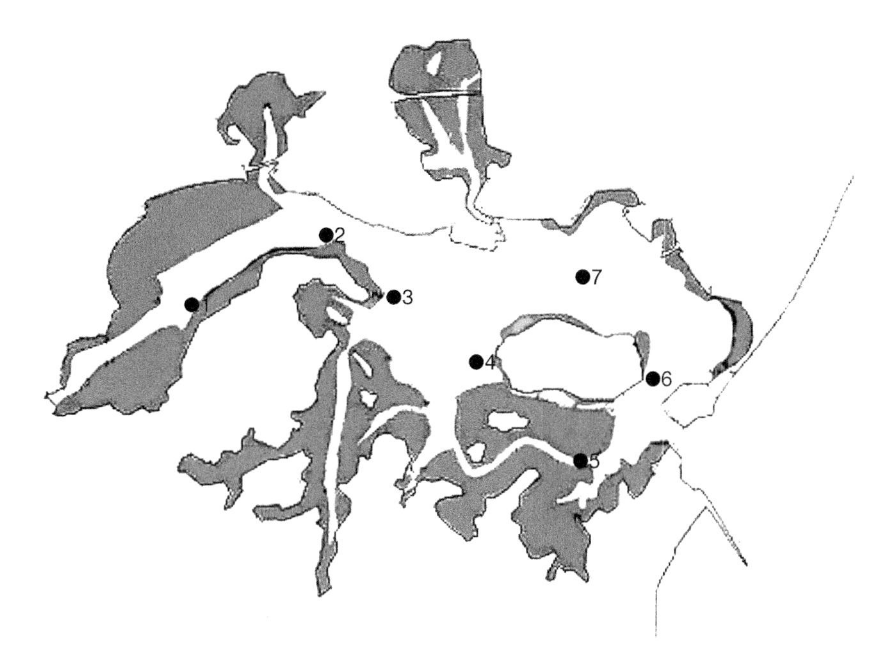

Figure 1 Sampling sites. Station 1: Wareham Channel (Buoy 84); Station 2: Wareham Channel (Buoy 74); Stations 3 and 4: Upper Wych Channel (Shipstal Point and Ramhorn Lake); Station 5: South Deep (between Furzey and Green Island, No. 13); Station 6: South Deep (Buoy 42); Station 7: Diver Middle Channel (Diver 51).

to achieve a structure which showed a degree of self-organization and predictability. This had a certain degree of independence from the physical environmental variables.

No obvious relationship was observed between the abundance of the major zooplanktonic groups and temperature, salinity or phaeo-pigments. It is suggested that the higher concentration of larval stages in the western part of the harbour compared with the eastern part of the system might be caused by a residual circulation and the more sheltered conditions of the harbour. These could promote an increase in primary production and larval retention. Thus, temporal change in abundance of species recorded in the west part of Poole Harbour, toward the end of the summer, appeared to be linked with the increase of chlorophyll 'a' concentration observed during late summer (Figure 3).

The influence of tidal import (and export) of neritic forms to the system from the coastal environment of Poole Bay must also be considered when looking at zooplankton composition at the surface. However, organization and temporal succession of zooplankton in the harbour suggested that the temporal abundance and structure of Poole Harbour zooplanktonic communities were mainly dependent on autochthonous production, food availability and tidal and residual circulation.

Table 1 Abundance of zooplankton groups (abundance as numbers per m^3)

Stn*	Calanoids	Copepod nauplii	Herpacticoids	Cirripede nauplii	Gastropod larvae	Bivalve larvae	Polychaete larvae	Decapod larvae
May	1808	17	0	626	122	35	52	0
1	1600	35	0	52	365	0	17	17
2	87	17	0	69	17	17	0	0
3	904	35	0	122	17	0	17	0
4	122	17	0	87	17	0	17	0
5	122	0	0	87	52	0	17	0
6	191	17	17	35	313	0	70	17
7								
June								
1	113	487	9	157	183	165	17	0
2	557	591	17	930	122	70	87	70
3	365	461	9	235	35	17	104	9
4	209	1017	35	313	304	78	104	52
5	435	1409	78	365	200	96	183	96
6	391	991	96	304	113	52	69	113
7	383	713	148	226	183	44	78	113
July								
1	2487	1478	35	313	696	70	156	0
2	278	191	70	557	174	104	52	0
3	748	661	174	504	87	209	122	0
4	1078	957	104	278	209	244	469	0
5	244	296	191	122	244	226	17	0

6	487	400	330	296	244	17	35	0
7	296	139	191	174	52	35	70	0
August								
1	1861	2574	87	348	87	104	70	0
2	278	765	209	626	52	383	87	17
3	313	1270	70	209	52	365	35	17
4	748	1270	417	174	87	539	157	0
5	226	661	313	104	104	226	52	0
6	278	574	626	70	139	104	70	0
7	35	313	278	52	52	104	52	0
September								
1	23478	8348	1130	435	174	435	4435	0
2	1426	1148	244	122	122	122	17	0
3	1426	2470	226	122	591	417	330	0
4	504	1757	348	157	261	400	70	0
5	87	1148	469	0.0	104	122	70	0
6	90	1126	469	0.0	104	122	70	0
7	557	800	469	87	35	348	87	0

* See Figure 1 for station positions.

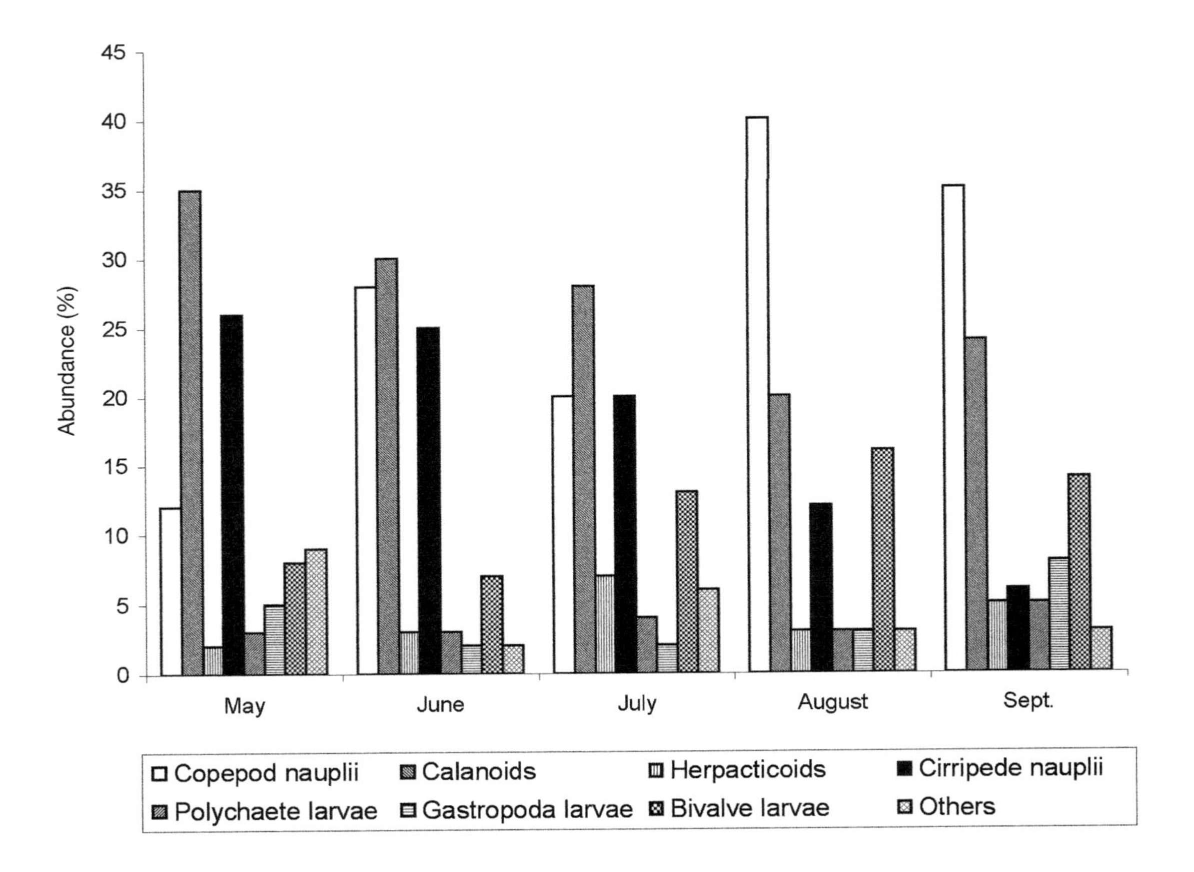

Figure 2 **Composition of the zooplankton community of Poole Harbour.**

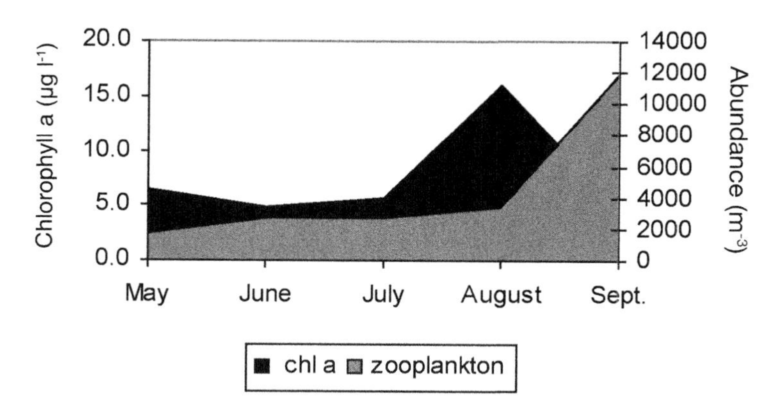

Figure 3 Chlorophyll 'a' concentration and zooplankton abundance.

References

Ambrogi, R., Ferrari, I., Geraci, S. and Ros, J. (1987) Biotic exchange between river, lagoon and sea: The case of zooplankton in the Po Delta. pp. 601–608. In: *European Marine Biology Symposium, Barcelona.*

Barbuto, P. C. (1999) A Preliminary and Ecological Survey of the Summer Macro and Mesozooplankton in Poole Harbour, An Open Coastal System, Dorset, South England. Dissertation submitted in partial fulfilment of the requirements for the degree of M.Sc. (Oceanography), University of Southampton.

Barnes, R. S. K. (1974) *Coastal Lagoons*. Cambridge: Cambridge University Press.

Burkill, P. H. (1983) Sampling and analysis of zooplankton populations. pp. 139–183. In: *Practical Procedures for Estuarine Studies.* Morris, A. W (ed.). London: NERC.

Collins, N. R. and Williams, R. (1981) Zooplankton of the British Channel and Severn Estuary. The distribution of four copepods in relation to salinity. *Marine Biology*, **64**: 273–283.

Collins, N. R. and Williams, R. (1982) Zooplankton communities in the Bristol Channel and Severn Estuary. *Marine Ecology Prog. Series* **9**: 1–11.

Colombo, G., Cecchierelli, V. and Ferari, I. (1984) Lo zooplankton delle lagune. *Nuova Thalassa,* **2**: 185–200.

Jouffre, D., Lam-Hoai, T., Millet, B. and Amanieu, M. (1991) Structuration spatiale des peuplements zooplanctoniques et fonctionnement hydrodynamique en milieu lagunaire. *Ocean. Acta*, **14**: 5.

Raymont, J. E. E. (1963) *Plankton and Productivity in the Oceans*. Oxford: Oxford University Press.

Wilson, J. G. (2002) Productivity, fisheries and aquaculture in temperate estuaries. *Estuary, Coastal Shelf Science*, **55**: 953–967.

10. The Important Birds of Poole Harbour: Population Changes Since 1998

Bryan Pickess

8 Shaw Drive, Sandford, Wareham, Dorset BH20 7BT

Details are given of the changes that have occurred since 1998 of the important birds of Poole Harbour. This 5 year survey included 10 breeding, a 'globally threatened' autumn passage migrant, and 20 international, national and harbour important wintering species. There are now over 40 breeding Little Egrets, the Black-headed Gull has been afforded 'Amber' status and concern is expressed for the survival of breeding Redshank in the harbour.

Peak Shelduck numbers have declined and it no longer qualifies for international status. Black-tailed Godwit numbers continue to rise and the Avocet is increasing rapidly: both populations are of international importance. Ten species are showing upward trends but seven are showing downward trends. Three species no longer occur in nationally important numbers. Only the numbers of the two small grebes and Oystercatcher appear to have changed little since the 1991/92–1997/98 period. The most significant declines have been noted with Pochard, Lapwing and Redshank.

Introduction

Poole Harbour has for more than four decades been acknowledged as an important national biological site by being afforded numerous statutory designations. Among these designations is the recognition of Poole Harbour as a wetland of international importance under the RAMSAR Convention and as a Special Protection Area (SPA), under the European Community Birds Directive, both ratified in 1999.

The report *Important Birds of Poole Harbour* (Pickess and Underhill-Day, 2002) gave the current status and historical data concerning the important breeding species up to 1998 and the passage and wintering waterfowl and waders to the winter of 1997/98. Five years have elapsed since that survey, so it is appropriate to review the status of the key breeding, passage and wintering species, and to record any changes that have occurred.

Breeding species

During the review period, there have been no national surveys involving the key harbour breeding species. Most of the breeding species in the report by Pickess and Underhill-Day (2002) were selected because they were included as 'Red' or 'Amber' species in Gibbons *et al.* (1996). Two additional species were added because the harbour held over

1% of the UK breeding population, although they were not listed as'Red' or 'Amber'. The breeding status of waterfowl and wader species, where peak numbers on passage or during the winter are of international or national importance, is also included.

A revised assessment of the status of birds of conservation concern for the period 2002–07, not surprisingly includes all but one of the harbour's previously listed 'Red' or 'Amber' species (Gibbons *et al.*, 1996). In addition, the Black-headed Gull *Larus ridibundus*, which was already recognized as an important harbour breeding bird, is now 'Amber' listed (Gregory *et al.*, 2002).

During the period 1998–2002, there was little detailed knowledge of how the harbour's key species fared. Table 1 shows these key species, their changing national status, if any, other important breeding species in Poole Harbour and as a percentage of the UK population.

Table 1 Changes in the status of important Poole Harbour breeding species between 1996 and 2002

Species	Red list		Amber list		Harbour importance		2002 % UK population
	1996	2002	1996	2002	1996	2002	
Little Egret[1]			■	■			Possibly >75%
Water Rail[2]			■	■			Unknown - <20 pairs
Redshank[4]			■	■			0.5%
Mediterranean Gull[1]			■	■			2-4%
Black-headed Gull[2]				■	■		>4%
Sandwich Tern[3]			■	■			>1.5%
Common Tern[5]					■	■	>1%
Cetti's Warbler[6]			■			■	>5%
Bearded Tit[7]			■	■			>2%
Reed Bunting[8]	■	■					Possibly >100 pairs

Source: After Gibbons *et al.* (1996); Gregory *et al.* (2002)

[1] 5 year mean of 1–300 breeding pairs in UK.
[2] Moderate (25–49%) contraction of UK breeding range over last 25 years.
[3] European listed SPEC2.
[4] 'Amber' status because of winter numbers but harbour also supports the largest number of breeding birds in south-west England.
Harbour breeding population >1% of UK population.
[6] No longer a rare breeder (>300 pairs).
[7] >50% of UK breeding population in 10 or fewer sites.
[8] Rapid (>50%) decline in UK breeding population over last 25 years.

Almost all of our current knowledge concerns the larger species which are easier to census. Undoubtedly, the highlight of the past 10 years has been the establishment by the Little Egret *Egretta garzetta*, on the Dorset Wildlife Trust's Brownsea Island nature reserve. Currently the breeding population has reached around 45 pairs, which represents over 75% of the UK breeding population. This colonization illustrates how rapidly changes can take place but not always in a positive way. Although the Grey Heron *Ardea cinerea* is not a threatened species, Brownsea Island once held one of the largest heronries in the UK, with up to 131 nests in 1971, but it declined to around 100 nests by the early 1980s (Prendergast and Boys, 1983). Numbers have continued to decline slowly and in 2003, there were only 35 nests (Chris Thain, pers. comm.).

One of the most difficult birds to census is the Water Rail *Rallus aquaticus* and until a concerted effort is made to undertake a breeding survey, its population size in the harbour remains unknown. Although it seems unlikely that the harbour population is of national significance, it is, nevertheless, an important member of the harbour's breeding bird community.

No recent surveys of breeding Redshank *Tringa totanus* have been undertaken but observations suggest a decline may be occurring, possibly due to habitat change and disturbance during nesting. This is due to the current high level of grazing pressure by the Sika Deer *Cervus nippon* on the saltings, and the possible disturbance to nesting birds, with potential trampling of nests and chicks. Herds of up to 100 animals have been seen roaming across areas such as Arne Bay (pers. obs.).

The Black-headed Gull, previously included among the harbour's most important breeding species because of its large population, has now been 'Amber' listed. Although the size of the colony on the *Spartina* islands in Holton Bay has not been estimated recently, there is no evidence to suggest a decline. Within the gull colony, Mediterranean Gulls *Larus melanocephalus* are still present, and their breeding population is probably between two and eight pairs. There has recently been an increase in sightings away from the nesting islands during the breeding season, which may suggest that the harbour population is slowly increasing.

The islands specially constructed by the Dorset Wildlife Trust to attract nesting terns on to their Brownsea Island lagoon had their best year ever in 2003, with 207 pairs of Common Tern *Sturna hirundo* and 194 pairs of Sandwich Tern *Sturna sandvicensis*. Sandwich Terns are fickle, however, for after many years of continuous breeding, for reasons unknown, they did not settle to nest in 2000 or 2001, but returned to nest again in 2002.

The Cetti's Warbler *Cettia cetti* was previously afforded 'Amber' status as a rare breeding bird but it has now been transferred to the 'Green' list due to its expanding population (Gregory *et al.*, 2002). Around Poole Harbour, it favours the wet fringes where the reed *Phragmites australis* is found, usually in association with scattered bushes of sallow *Salix* sp. and patches of bramble *Rubus* sp. The harbour still supports over 5% of the national population and it remains a key breeding species. Whilst the UK breeding population is currently expanding, the significance of the harbour's population will only be tested when the next severe winter occurs. Cetti's Warbler is vulnerable during a long period of severe weather, which can lead to local extinctions (Batten *et al.*, 1990). During such times the harbour, which usually experiences slightly milder conditions than the other main breeding areas in South East England and East Anglia, can provide conditions which allow higher survival rates during cold weather.

Often found in a similar habitat to the Cetti's Warbler, the Reed Bunting *Emberiza schoeniclus*, has suffered a drastic population decline over the past 25 years and is 'Red' listed. They are still thinly distributed around the southern and western fringes of the harbour but an estimated breeding population in excess of 100 pairs may be optimistic. There is a need to survey this species, the only 'Red' listed breeding species nesting in the harbour. Associated with the larger reedbeds on the western and north-western side of the harbour, a small population of Bearded Tit *Panurus biarmicus* is at the most 10 pairs. Although numbers are small, it represents over 2% of the UK breeding population.

A 'Red' listed autumn passage passerine

The Aquatic Warbler *Acrocephalus paludicola*, a globally threatened and 'Red' listed non-breeding migrant species, probably passes through the harbour annually each autumn, but its presence is usually only detected through mist netting for other species. During the 1990s, it was recorded regularly up to 1997, with a maximum of 17 in any one autumn at the Keysworth trapping site (Pickess and Underhill-Day, 2002). Subsequently, birds have been recorded (caught) in only one of the past 4 years, when in August 2000, three were trapped on 5th and one on 10th. The recent fall in the number of birds in Poole Harbour is consistent with other sites along the Dorset coast, where records of this species have also declined since 2000, their best year.

Passage and wintering

The low water WeBS counts have continued each month, September–March, in each of the 5 years 1998/98–2002/03. There has also been a revision of the national 1% thresholds for wintering waterbirds (Kershaw and Cranswick, 2003) and waders (Rehfisch *et al.*, 2003), and the 1% thresholds for international waterbirds (Delany and Scott, 2002).

During 1991/92–1997/98, 21 species met the criteria for inclusion by Pickess and Underhill-Day (2002), as they occurred in the harbour in internationally important numbers (2 species), nationally important numbers (16 species) or their peak population represented over 5% of the harbour's total peak wintering population of waterbirds and waders (3 species).

With the exception of the Grey Plover *Pluvialis equatarola,* which no longer meets any of the criteria for inclusion, the remaining 20 species still meet at least one of the criteria for inclusion, although the status of several has changed (Table 2). The indication of population trends in the harbour for each species has been calculated by using the annual peak counts for the 11 winters 1991/92–2001/02. The same data set has been used, in text, to calculate the 11 year mean for each of the harbour's wintering waterbirds and waders of importance.

Currently at peak, two species, both of which are increasing in numbers, occur in internationally important numbers in the harbour, the Avocet *Recurvirostra avosetta* and Black-tailed Godwit *Limosa limosa islandica.* Although the 1% threshold has not changed for Shelduck *Tadorna tadorna,* a decline in the number of this species means that it no longer qualifies for international status.

The most significant change since the winter of 1997/98 has been the Avocet, as not only have its numbers increased to reach international importance but during 2001/02 a peak count of 1862 was also recorded, the largest gathering ever at one site in the UK. Until the end of the 1970s, this was a rare winter visitor to the harbour and its rapid rise during the 1990s to the current time can only be described as meteoric. Two pairs attempted to breed in 1999 on the lagoon of Brownsea Island nature reserve and, although the eggs hatched, the young chicks soon disappeared: the cause of their demise is unknown.

The wintering Black-tailed Godwit population of the harbour has been of international importance for many years. Over time, at peak, the population has slowly but steadily increased and during this review period, it has peaked at around or just over the 2000 mark, and the 11 year mean is 1668. A record count of 2115 in 2002 represented over 14% of the UK wintering population and 6% of the East Atlantic Flyway/north-west European population of this species (Rehfisch *et al.*, 2003).

The peak winter Shelduck populations during the six winter periods 1991/92–1996/97 produced a mean of 3175, the 1% international threshold being 3000 birds. From 1997/98, the population has dropped by over 30% at peak and the harbour no longer holds internationally important numbers, even though the 11 year mean is 3052. What has caused this rapid drop is unclear but it is not just a Poole Harbour phenomenon, as the UK winter population overall has also been falling (Pollitt *et al.,* 2003).

Always difficult to census are the Slavonian Grebe *Podiceps auritus* and Black-necked Grebe *P. nigricollis.* The recent records of Black-necked Grebe suggest that at its maximum it is failing to reach double figures, whereas the Slavonian Grebe is faring better, with a peak of 15 birds. At peak populations, both grebes are over 1% of the UK population.

Only 7 of the 14 nationally or harbour important species show upward trends. There is no surprise in the rising fortunes of the Cormorant *Phalacrocorax carbo* in the harbour. Despite >75% rise in the 1% national threshold to 230, the harbour now holds nearly 2% of UK winter population, and the 11 year mean is 388. The recent breeding colonist, the Little Egret, has an increasing wintering population in the harbour and reached a peak of

Table 2 Changes in the status of key wintering species and their trends over the period 1991/92–2001/02 (11 years) in Poole Harbour

Species	1991/92–1997/98	1998/89–2002/03	Current trends
Slavonian Grebe	NI	NI	Population very small – probably stable
Black-necked Grebe	NI	NI	Population very small – probably stable
Cormorant	NI	NI	Increasing
Little Egret	NI	NI	Increasing and 'Amber' listed as a breeding species
Dark-breasted Brent Goose	NI	NI	Declining
Shelduck	IN	NI	Declining
Wigeon	H	H	Slowly declining and below 1% threshold
Teal	NI	NI	Increasing but now below 1% threshold
Pintail	NI	NI	Increasing
Shoveler	NI	H	Declining and now below 1% threshold
Pochard	NI	H	Declining and now below 1% threshold
Goldeneye	NI	H	Slowly increasing but now below 1% threshold
Red-breasted Merganser	NI	NI	Increasing
Oystercatcher	H	H	Stable
Avocet	NI	IN	Increasing
Lapwing	H	H	Rapid decline and below 1% threshold
Dunlin	NI	NI	Increasing
Black-tailed Godwit	IN	IN	Increasing
Curlew	NI	NI	Decreasing
Redshank	NI	H	Declining and now below 1% threshold

IN = International Importance
NI = National Importance
H = Harbour Importance

197 during the winter of 2001/02, with the 11 year mean of 65.

Numbers of Teal *Anas crecca* are slowly increasing in the harbour but the increase of the 1% threshold to 1920, now places the Teal below the qualifying figure. The 11 year mean is 1487. Similarly, there has been a slow but steady increase in the numbers of Pintail *Anas acuta,* with the 1% national threshold now at 279 and the 11 year mean of 294. The trend for Goldeneye *Bucephala clangula* suggests a slow rise but with the 1% threshold increasing to 249, only in 2 of the past 11 years would this new qualifying level have been reached: the 11 year mean is 217. Furthermore, the population had by 2001/02 declined to near its 1991/92 peak of only 122. The harbour is the most important estuary in England for wintering Red-breasted Merganser *Mergus serrator*, and numbers are slowly moving upwards. At the 1% national threshold of 100 and the 11 year mean of 391, the harbour holds around 4% of UK wintering population.

The only wader to show an upward trend is the Dunlin *Calidris alpina* and even with the 1% national threshold being raised by 6% to 5600, the 11 year mean at 6134 is well above the qualifying number.

The only species whose numbers are stable is the Oystercatcher *Haematopus outrageous* which, despite an 11% drop in the 1% national threshold to 3200, remains only a harbour important species with an 11 year mean of 1580.

Of the species in decline, only the Dark-breasted Brent Goose *Branta bernicula* and Curlew *Numenius arquata* numbers remain above the 1% threshold. Three species, Shoveler *Anas clypeata*, Pochard *Aythya ferina* and Redshank, were all previously above the 1% national threshold, but no longer qualify and are relegated to harbour important status. The Wigeon *Anas penelope* and Lapwing *Vanellus vanillas,* both previously with populations of harbour importance, have declined.

During the 1990s, peak numbers of Dark-breasted Brent Goose have varied but the trend has been one of slow decline, although numbers have always been well above the 1% national threshold (1000) and the 11 year mean of 1409. The winter of 2001/02 saw a drastic decline with a peak count of only 578. The reason for this poor showing is unclear. The 1% national threshold for Curlew has been revised upwards by 25% and is now 1500, whilst the harbour 11 year mean is 1687. Although the trend for this species in the harbour is slowly downward, the annual peaks remain around or well above the new 1% threshold level.

The Shoveler has never been an abundant bird in the harbour and populations have been given to large fluctuations. Despite reaching the new 1% national threshold of 150 in 6 of the last 11 years, it nevertheless only has an 11 year mean of 114. The new 1% national threshold of 595 for Pochard has risen sharply, whilst the numbers in the harbour in recent winters have declined rapidly. The 11 year mean of 602 is only due to three exceptionally large counts before 1998/99. Unless numbers visiting the harbour increase again, it is likely to be downgraded to harbour important. A real cause for concern is the Redshank, where the trend has been steadily downward and during the winters of 2000/01–2001/02, the population peaked at just below 800. The rapid drop in wintering birds has reduced the 11 year mean to 1171, and is now below the 1% national threshold, recently raised from 1100 to 1200.

Two 1991/92–1997/98 harbour important species have shown downward trends. Winter numbers of Wigeon have always fluctuated but the recent trend has been downward and the 11 year mean is 1006. Lapwing numbers have rapidly declined from peaks of >8000–10,000 in the early 1990s to just over 600 during the winter of 2001/02: the 11 year mean is 4045.

Discussion

The distribution and status of the breeding species in Poole Harbour appear to show little change during the past 5 years. Because of the nationally declining breeding population of the Black-headed Gull, this species is now 'Amber' listed. It would be valuable to establish around the harbour the breeding status of the Water Rail and more importantly, our only 'Red' listed species, the Reed Bunting. The need to census the Redshank is more urgent to discover if the large herds of Sika Deer are having a serious impact upon breeding success.

What is actually happening with the harbour's wintering populations is not clear. Could one of the reasons for these recent changes be the series of milder winters that have taken place in the UK and north-west Europe? It is unlikely to be quite this simple because some species are showing downward trends, whilst a smaller number are steadily increasing. Should these trends continue, is the importance status of several species in the harbour likely to change?

Acknowledgements

I would like to thank John Day for commenting on the first draft and making some very helpful comments.

References

Batten, L. A., Bibby, C. J., Clement, P., Elliott, C. D. and Porter, R. F. (eds) (1990) Cetti's Warbler. In: *Red Data Birds in Britain*. London: Poyser.

Delany, S. and Scott, D. (2002) *Waterbird Population Estimates*. Third Edition. *Wetlands International Global Series,* No. 12. Wageningen.

Gibbons, D., Avery, M., Baillie, S., Gregory, R., Kirby, J., Porter, R., Tucker, G. and Williams, G. (1996) Bird species of conservation concern in the United Kingdom, Channel Islands and Isle of Man: revising the Red Data List. *RSPB Conservation Review*, **10**: 7–18.

Gregory, R. D., Wilkinson, N. I., Noble, D. G., Robinson, J. A., Brown, A. F., Hughes, J., Procter, D., Gibbons, D. W. and Galbraith, C. A. (2002. The population status of birds in the United Kingdom, Channel Islands and Isle of Man: an analysis of conservation concern 2002–2007. *British Birds,* **95**: 410–448.

Kershaw, M. and Cranswick, P. A. (2003) Numbers of wintering waterbirds in Great Britain, 1994/1995–1998/1999: I. Wildfowl and selected waterbirds. *Biological Conservation,* **111**: 91–104.

Pickess, B. P. and Underhill-Day, J. C. (2002) *Important Birds of Poole Harbour*. Wareham: Poole Harbour Study Group.

Pollitt, M., Hall, C., Holloway, S., Hearn, R., Marshall, P., Musgrove, A., Robinson, J. and Cranswick, P. (2003) *The Wetland Bird Survey 2000–01: Wildfowl and Wader Counts.* Slimbridge BTO/WWT/RSPB/JNCC.

Prendergast, E .D. V. and Boys, J. V. (1983) *The Birds of Dorset.* Newton Abbot: David and Charles.

Rehfisch, M. M., Austin, G. E., Armitage, M. J. S., Atkinson. P. W., Holloway. S. J., Musgrove, A. J. and Pollitt, M. S. (2003) Numbers of wintering waterbirds in Great Britain and the Isle of Man (1994/1995–1998/1999): II. Coastal waders (Charadrii). *Biological Conservation,* **112**: 329–341.

11. Otters in Poole Harbour

Bronwen Bruce

Dorset Wildlife Trust, Forston Farm, Brooklands, Dorset DT4 0AP

Historical evidence suggests that Otters used Poole Harbour extensively in the past and today there is a range of habitat features that should provide opportunities for a thriving Otter population. However, records collected by the Dorset Otter Group since 1997 have been scarce and irregular, suggesting that Otters have not colonized the harbour to the extent that they did in the 1960s before the population underwent a dramatic crash. A lack of information restricts full understanding of the potential barriers to Otters colonizing the harbour and the possible solutions. Initial actions should, therefore, focus on surveying and research. However, other problems could include Otters drowning in unregulated fyke nets which are used to catch eels, lack of holt or resting sites close to freshwater, increasing boat traffic, development and roads, potential pollution problems and possible declines in fish stocks.

Introduction

A decline in the population of European Otter *Lutra lutra* in Dorset in the late 1950s continued into the 1980s with the species ceasing to occupy previous territories including Poole Harbour. This decline is believed to be the result of pollution of watercourses from pesticides, particularly organophosphates, and through loss of habitat. In 1978, the Otter became legally protected in England and Wales and various conservation bodies worked to reverse the declining population.

In 1995, the Otter was identified as a 'Priority' species for action in the UK Biodiversity Action Plan (JNCC, 1995). The objectives of the national Otter Biodiversity Action Plan (BAP) are:

- to maintain and expand existing Otter populations
- by 2010, to restore breeding Otters on all catchments and coastal areas where they have been recorded since 1960.

Historical context

Poole Harbour should offer excellent opportunities for Otters as there is an abundance of prey items and suitable habitat within the harbour. The name of 'Otter Island' in Lytchett Bay suggests that Otters were a feature of the harbour in the past. However, we need to examine historical records to check that the harbour did naturally support Otter populations.

The Courtenay Tracy Otter Hunt took over informal hunts in Dorset in 1887 on most rivers, but formal records for Otter kills were not kept until the 1930s. However, although there are plenty of records for Otters on rivers flowing into the harbour, there are few for Poole Harbour itself as not many hunts occurred here. During this period, the majority of records come from naturalists or school natural history societies such as Bryanston and Clayesmore. As the BAP target relates to the Otter population in the 1960s, records from this time are particularly important. These are shown in Table 1 and the distribution of these records is illustrated in Figure 1.

Many of the records for Poole Harbour between 1960 and 1979 lack detail. Only the few Otter hunt records give information on what the observation was and most records are four figure grid references so exact locations are not too clear. Yet they do indicate that Otters were widespread in Poole Harbour up until the 1970s when records become scarcer. In order to fulfil the BAP objectives, therefore, we need to reach a stage when Otters are recorded as widespread again.

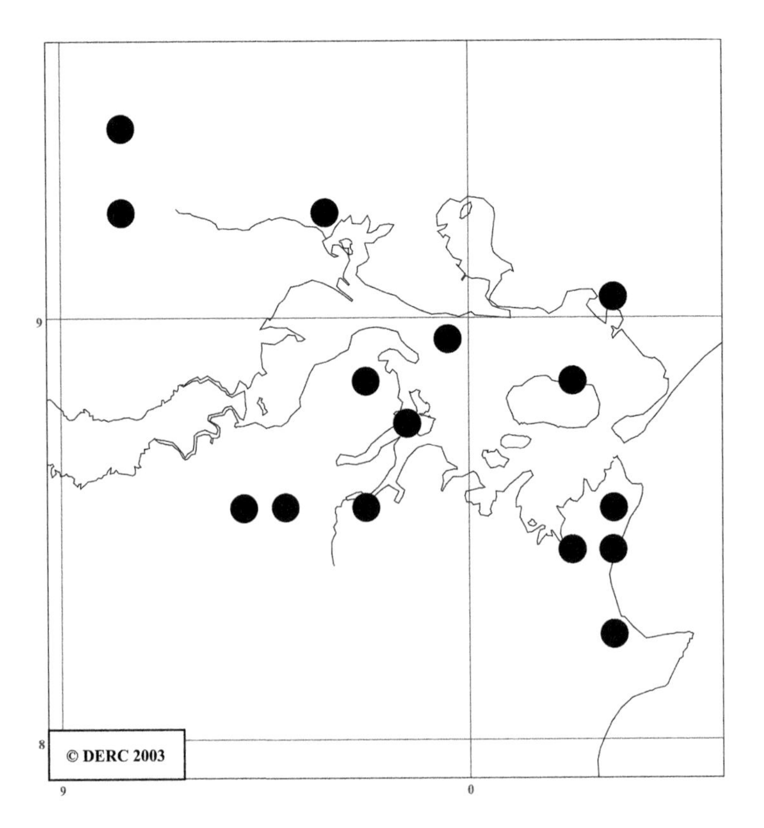

Figure 1 Records for Otters in Poole Harbour, 1960–79.

Table 1 Otter records in Poole Harbour 1960–79

Grid reference	Date	Recorder	Area in Poole Harbour	Observation
SY961929	26/7/61	Courtenay Tracy Otter Hunt	Lytchett Minster, footbridge	2 Otters (1 male 20 lb)
SZ0384	7/2/63	Brotherton, H.	Studland Beach	Unknown
SY978856	8/4/63	Courtenay Tracy Otter Hunt	Wytch Farm, far side of estuary	Otter 17 lb
SZ0382	29/3/64	Unknown	Studland NNR	Unknown
SY961929	26/9/64	Courtenay Tracy Otter Hunt	Lytchett Minster	Drag
SZ0288	2/65	Bromby, A.	Brownsea Island	Unknown
SZ0288	5/65	Bromby, A.	Brownsea Island, West Lake	Unknown
SY9585	1967	Duff, K.	Slepe Heath	Unknown
SY9485	1967	Duff, K.	Hartland	Unknown
SY9192	6/67	Cox, J.	Sherford River	Unknown
SZ0384	22/9/67	Teagle, W.	Studland	Unknown
SZ0385	2/12/67	Teagle, W.	Studland	Unknown
SZ0385	3/12/67	Teagle, W.	Studland	Unknown
SZ0385	22/12/7	Teagle, W.	Studland	Unknown
SZ0284	1968	Cox, J.	Littlesea	Unknown
SZ0385	1968	Teagle, W.	Studland	Unknown
SZ0288	1968	Wise, A.	Brownsea Island	Unknown
SY977887	1969	Pickess, B.	Arne	Unknown
SZ0390	12/69	Brotherton, H.	Poole	Unknown
SZ0390	1/70	Brotherton, H.	Poole Harbour	Unknown
SZ0284	1970	Unknown	Littlesea	Unknown
SZ0284	1973	Cox, J.	Littlesea	Unknown
SY9194	1973	Anon	Sherford River, Morden	Unknown
SY916926	26/7/73	Cox, J.	Sherford River, nr Morden	Unknown
SY9989	26/8/73	Milton Abbey School, Natural History Society	Poole Harbour, nr Wareham Channel	Unknown
SY9887	7/12/75	Prendergast, E.	Poole Harbour, nr Arne peninsula	Unknown

Source: Collated by James (1999).

Recent records

In 1997, the Dorset Otter Group, co-ordinated by the Dorset Wildlife Trust, was formed. The Group trains volunteers to look for signs of Otters throughout the county and has found a significant increase in these signs as well as recording an increase in the population on several rivers and in Christchurch Harbour. However, records for Poole Harbour have been scarce and inconstant. For example, the record for Brownsea in Table 2 indicates that an Otter was leaving regular spraint makings for a few months, but this did not continue unlike many Otter sprainting spots in the county. Continued regular sprainting in a specific spot indicates ongoing presence, and hence a territory. Sprainting for a short period may indicate an Otter investigating an area then moving on, unless another detrimental factor has occurred, such as a fatality.

The records in Table 2 are from the northern side of the harbour only, despite the southern side offering some of the best habitat opportunities for Otters. Moreover, of the four rivers flowing into Poole Harbour, only the Corfe River has no records for Otters despite surveying effort to detect Otter signs in what is a relatively accessible area.

These records suggest that Poole Harbour is not being occupied by Otters as regularly as the 1960s when regular records were gained from a variety of areas within the harbour. Despite checking of obvious sprainting/marking areas, such as on Brownsea, no current regular sprainting area has been identified within the harbour.

Table 2 Otter records in Poole Harbour

Grid reference	Date	Recorder	Area	Observation
SY9087	Spring 1999–Summer 2003	Lamming, J.	River Piddle, close to harbour	Spraint
SY9187	Summer 2000	Lamming, J.	River Piddle, close to harbour	Spraint
SY9388	Summer 2000	Lamming, J.	River Piddle, close to harbour	Spraint
SY9288	Spring 1999–Summer 2003	Windsor, S. and G.	River Piddle, close to harbour	Spraint
SZ0288	August 1999	Ramsey, J.	Brownsea Lake	Sighting
SZ0288	August 1999	Williams, C.	Brownsea boardwalk	Spraint
SY9192	Spring 2002	Fortesque, P.	Sherford Bridge	Spraint
SY9689	Summer 2003	Casual	Off Shag Looe Head	Sighting
SZ0089	Spring 2003	Kingston, S.	Lower Hamworthy	Dead in fyke net

Source: Provided by the Dorset Otter Group.

Habitats

There are a range of habitat features within Poole Harbour which can be utilized by Otters.

- Rivers – these are the habitat feature most commonly associated with Otters. Four main rivers flow into the harbour as well as the small Studland streams. All provide suitable habitat and the Rivers Frome and Piddle, where there are regular records for Otters, offer potential colonizing routes for Otters into the harbour.
- Sheltered bays – bays offer great opportunities for Otters. However, how well an Otter can exploit bays and harbours is related to the geology and landscape of the area. A coastal area that has freshwater pools, gentle shorelines and native woodland offers greater potential to Otters than a coastal area with high cliffs and intensive agriculture (IOSF, 1998). Poole Harbour is similar to the former and should be ideal for Otters. Prey density is often greater in marine areas than rivers and Otters can exploit whatever is available. Eels can still be found in the harbour and also flat fish, small fish such as sea scorpions or rocklings, molluscs and crustaceans. Some studies (Krunk *et al.*, 1998) have recorded Otters feeding entirely on items from the sea rather than freshwater habitats.
- Freshwater – if Otters are exploiting coastal areas then access to freshwater on a daily basis is essential for maintenance of their fur to preserve its thermo-insulating properties and to drink for maintenance of kidney function. There is a good supply of freshwater around Poole Harbour in both the lakes and rivers. Small pools and streams around coastal areas tend not to be used for foraging but purely for washing (Krunk *et al.*, 1998).
- Lakes – the two freshwater lakes on Brownsea and Littlesea Lake at Studland are available for Otters to wash in but are probably less important as food sources. These are oligotrophic lakes and have a low productivity, supporting only minnows and eels. Although these are popular food items for Otters, it is unlikely that a great amount of hunting effort would be focused here as there are much higher prey availability densities elsewhere in the harbour.
- Wet woodland – this offers both cover and food availability for Otters. Alder carr exists on Brownsea and along some of the Studland streams. Old alder stools that are above the waterline can provide potential holt sites for Otters.
- Reedbeds – offer prime habitat for Otters. The reed cover is extremely dense and Otters often do not require the covered holt sites used in other habitats. There have even been examples of females with young cubs raising them above ground in the centre of reedbeds on areas of raised ground. Reedbeds have good food availability, they are breeding areas for some coarse fish species, juvenile fish use reedbeds as refuge habitat, eel density tends to be high and amphibians can exploit the habitat. Some radio-tracking studies (Krunk *et al.*, 1998) showed a preference by Otters for reedbed and island habitats in freshwater habitats but it is unclear just how important they are in marine habitats.

Otter territories are large. Males can encompass 15 miles or more, so one territory can include a number of these habitat features, providing wide opportunities for Otters.

Problems

As habitat is good, Otter colonizing routes from existing populations are present and the historical records indicate past use of Poole Harbour, why are there not more records for Poole Harbour? One reason is the difficulty in surveying for Otters in the harbour area. Much of the surrounding habitat is inaccessible with deep water in reed and bog habitats: ideal for Otters, but unsuitable for Otter surveyors. This is particularly true on the southern side of the harbour where there are fewer records but probably the most habitat potential for Otters.

However, there are certain 'spots' where Otter signs are expected but which are not giving regular, positive results. Otters prefer unvegetated and often artificial structures for sprainting on, such as concrete, boardwalks, weirs and beneath bridges. Sites such as Brownsea Island offer structures such as these amongst ideal habitat. These structures are surveyed on a regular basis for Otter spraint but are not resulting in regular records. From the few records we have, there is an indication of some Otter activity. Yet, for a fully established Otter population to be recognized in Poole Harbour, there should be more regular records indicating a breeding population with defined territories rather than Otters passing through the harbour with undefined territories.

It is possible that there could be a factor or factors limiting the Otter population in Poole Harbour. Possibilities include

- Fyke nets – Poole Harbour is one of the most important eel fisheries in the country and fyke nets are used to catch eels. If these are registered by the Environment Agency, they should have coloured tags attached to indicate the year of registration and be fitted with an Otter guard. There is a legal requirement to fit an Otter guard under a national by-law. Fyke nets include a series of concentric circles as shown in Figure 2. Otters go into the net after eels but become trapped and drown within the net. An Otter guard stitched into the entrance of the net as shown in Figure 2 prevents an Otter from entering. On 6 March 2003, a male Otter was found dead in a fyke net at Lower Hamworthy by the Poole Harbour Commissioners. This was an unregistered net used illegally, and highlights the problem that exists when poaching occurs.
- Declining prey items – various fish stocks are believed to be declining, including eel stocks. If prey items are not as abundant as in the 1960s, it is possible that this could limit the Otter population size. However, with eels still being caught in good numbers in Poole Harbour, it seems unlikely that this would stop them from colonizing altogether. Other items such as sea scorpions and flat fish are also recorded within the harbour so it appears that food is available.

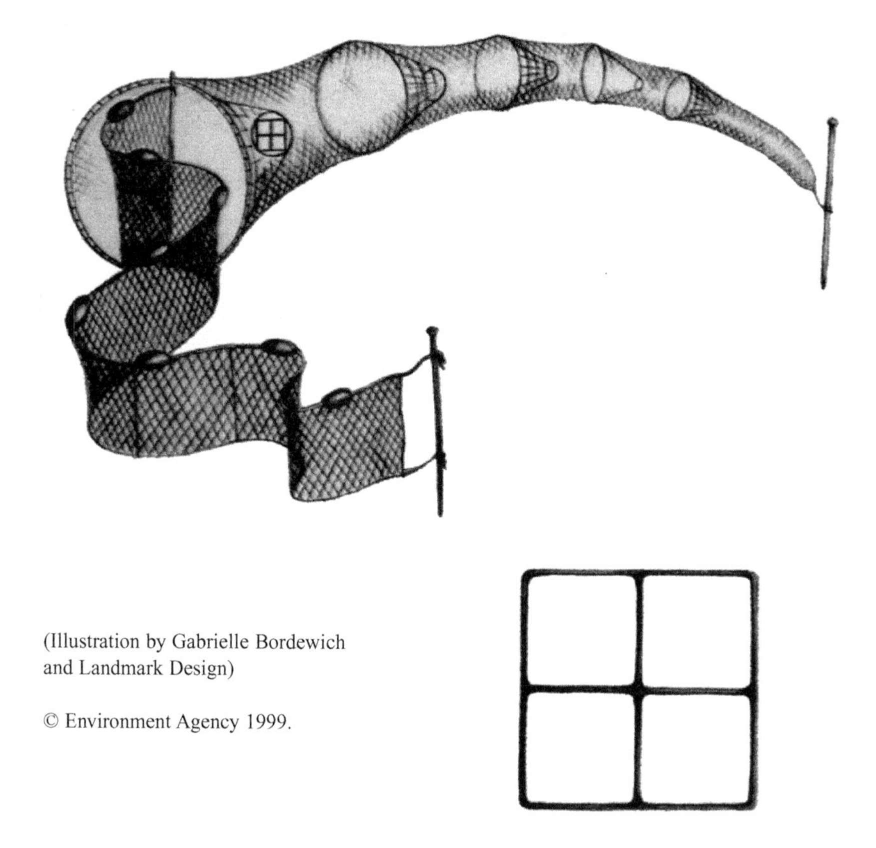

(Illustration by Gabrielle Bordewich and Landmark Design)

Figure 2 Fyke net and Otter guard.

- Availability of holt sites – Otter holts may be more important in coastal areas than freshwater as, despite washing in freshwater daily, it has been suggested that coastal Otters may have a greater problem with thermal regulation than freshwater Otters. Otters may require covered holts to keep warm in coastal habitats in a way they do not in freshwater areas. Moreover, Krunk *et al.* (1998) have shown that coastal Otters which utilized rocky holes as well as tree roots for holts chose holt sites that were very close to a source of freshwater. In fact, small freshwater pools were often found within the holt itself.
- Reedbeds could be important as Otters have made couches (above ground resting sites as opposed to covered holts) within reedbeds. These are quite elaborate ball-shaped designs made from bitten-off reeds. It is thought that these offer some insulating properties to help with thermal regulation. However, most reedbeds in Poole Harbour are saline so this may limit Otters' ability to exist in these areas and restrict resting sites to freshwater reedbed areas.

- Development and roads – there has been a large increase in development around Poole Harbour, but Brownsea Island and the Purbeck side of Poole Harbour have been largely unaffected. The Holes Bay and Sandbanks areas have seen a significant increase in development leading to the loss of some harbour edge habitat on the Poole side. Traffic on roads has significantly increased since the 1960s, which can cause problems throughout Dorset. In 1999, an unconfirmed report of a dead Otter was received on the A35 near Holes Bay. Unfortunately, a body was not picked up to confirm the report but it coincided with the disappearance of the Otter from Brownsea Island. We cannot be sure that these incidents are linked, and there is the potential for Poole Harbour to contain many Otters, but it could also have been the only or one of few Otters in the harbour thus greatly affecting the population.
- Boat traffic – the amount of boat traffic in the harbour has expanded since the 1960s. This could disturb Otters, particularly if it occurs at night or could harm/kill Otters if they are hit by a boat. Many large boats enter the harbour and there has been an increase in marinas and sailing schools contributing to the rise in development around the harbour edge.
- Pollution – increased run-off from development, chemical treatments on boats and sewage have all contributed to pollution within the harbour, some of which is retained in the sediments over decades. In their position at the top of the food chain, Otters are quite sensitive to pollution, which was the primary reason for their decline in the 1950–70s. Pollution could be a problem within Poole Harbour if it affects the Otter's food items or causes isolated incidents with localized effects. However, there are no data to determine whether or not this is a problem.

Solutions

To restore the Otters to a viable breeding population in Poole Harbour, we need to determine which of the potential problems are limiting factors and if possible address these problems.

Surveying and research – a pressing need and more regular surveys for Otters in Poole Harbour are required. Landing stages and slipways, particularly on the islands, could serve as Otter sprainting sites and sailing clubs could be enlisted to examine such areas. Other user groups such as bird watchers, canoeists and fishing clubs could be requested to assist. One canoeing club has been trained to look for Otter signs, as this sport allows access to the more inaccessible areas of the harbour that most Dorset Otter Group volunteers cannot reach. Poole Harbour Study Group could be helpful in stimulating research and surveying within the harbour. Student projects could be designed to examine details on Otter behaviour.

If any Otter carcasses are found it is vital that these are sent for post-mortem to detect if any toxic substances are present in the tissues, if any intraspecific aggression has occurred (which could indicate that other Otters are around and defending territories) and, in the case of females, if any breeding has happened.

With regard to regulation – legislation exists which affects the Otter population in Poole Harbour, notably the requirement to fit an Otter guard into a fyke net under a national by-law. Although a conviction for breaking this law and failing to register fyke nets results in a £500 fine plus costs, it is unclear whether this is a big enough deterrent and whether fines should be increased in the future. The Otter is a listed species requiring special protection measures under the European Habitats Directive (92/43/EEC) and is listed in Annex 2 requiring the designation of Special Areas of Conservation (SACs) for sites supporting important Otter populations (DEFRA, 2003). However, the Otter is not a listed species for the Poole Harbour candidate SAC. This means that consideration of the Otter population is not required when determining whether the SAC is in a favourable condition, but there is a need consider the Otter under its own legal requirements. This requires attention in the future.

Also useful are schemes to benefit or ensure the protection of both freshwater or coastal wetland habitat. These will aid the Otter population, particularly if they include Otter holts. Artificial holts may well help increase the density of Otter populations within the harbour, but these have to be created alongside freshwater areas or with freshwater washing pools constructed next to any holts. Management of the freshwater reedbeds to encourage the Otter population could also be important.

Finally as some roads are very close to where Otters could be using the harbour, particularly at Holes Bay, mitigation may be required to reduce Otter casualties. Survey work is required to identify potential areas where Otters could have a problem with the road and suggest works that could mitigate against the effects.

References

DEFRA (2003) *European Protected Species Guidance Note*. London: Department for Food And Rural Affairs.

IOSF (1998) *Geology and Otter*. Skye: International Otter Survival Fund.

James, Z. (1999) *Dorset Otter Records 1700 –1999*. Brooklands: Dorset Wildlife Trust.

JNCC (1996) *Biodiversity: The UK Steering Group Report*. Peterborough: Joint Nature Conservation Committee.

Krunk, H., Carss, D. N., Conroy, J. W. H. and Gaywood, M. J. (1998) *Habitat Use and Conservation of Otters (Lutra lutra) in Britain: A Review*. Cambridge: Cambridge University Press.

12. Non-native Species in and around Poole Harbour

John Underhill-Day[1] and Peter Dyrynda[2]

[1]Royal Society for the Protection of Birds, Syldata, Arne, Wareham, Dorset BH20 5BJ

[2]School of Biological Sciences, University of Wales Swansea, Singleton Park, Swansea, West Glamorgan SA2 8PP

Poole Harbour and its environs have become home to a range of alien aquatic and terrestrial flora and fauna. Some, like the hybrid Cord Grass and the Sika Deer have had a major impact on the habitats of the area, while others seem to have had little effect. This chapter describes the wide range of species, including those of the marine environment which have arrived or been introduced to the area in recent years.

Introduction

The issue of aliens and introductions is a vexed one. The introduction of non-native species either by accident or design is recognized as one of the major causes of the loss of biodiversity on our planet (IUCN, 2000). However, in Britain, no endemic plants are thought to have been made extinct, and there appear to be few rare plants threatened by introductions. If we include garden species, the British Isles hosts some 12,500 vascular plant species, of which only some 1400 are native, however, only 14 of this huge number of introduced species have become serious pests (Young, 2000).

In recent years, the rate at which introductions have arrived seems to have quickened. For example, one new aquatic vascular plant species entered the British flora in the 50 years up to 1850. In the following 50 years, the number was four, during the next 50 years, seven, and between 1950 and 2000, nine new species arrived. In 1928, Druce (1928) listed 293 vascular plant species as established aliens in Britain, an introduction rate of probably about one species a year over the previous 300 years. The latest lists, published in 1992 and 1994, list 889 established aliens, an average increase of about nine a year since Druce (Kent, 1992; Clement and Foster, 1994).

Introduced mammals are fewer in species, but can do a disproportionate amount of harm. Although, in Britain, the introduced North American Muskrat *Ondatra zibethicus* and the South American Coypu *Myocaster coypus* were successfully eliminated in 1937 and 1989, respectively, escaped Mink *Mustela vision,* and introduced Grey Squirrels *Sciurus carolinensis,* still pose a serious threat, the first to our native Water Vole *Arvicola terrestris,* and the other to the Red Squirrel *Sciurus vulgaris* populations, and both these introduced species seem unstoppable.

The Poole Harbour situation

The Poole Harbour basin has not been immune from these processes. The three mammals most likely to be seen are the Rabbit, introduced by the Normans in the twelfth century, the Grey Squirrel, first brought into Cheshire from North America in 1876, and then followed by another 30 introductions across England and Wales up to 1929, and the Sika Deer *Cervus nippon* introduced on to Brownsea Island from Japan in 1896.

Many of the drier woodland areas around the harbour contain Sycamore *Acer pseudoplatanus*, probably introduced into Britain in the fifteenth century and now found in 11 of the 19 woodland communities in Britain listed in the National Vegetation Classification (Rodwell, 1991). Areas of acid woodland and heath often contain *Rhododendron ponticum*, which has been here since at least 1763 and is one of the serious pests referred to earlier. On the Arne peninsula to the south of the harbour, the elimination of Rhododendron in the next 5 years is finally contemplated, a process which began in the mid-1960s.

In contrast, two species, *Cotoneaster simonsii,* found growing in woods and on heaths and roadsides around the harbour, and a more recent introduction, the Mexican Daisy *Eiogeron karvinskianus*, found on walls and in paved areas around Wareham, seem to offer little threat to native species. The former is one of 67 species of *Cotoneaster* growing in the wild in Britain, the commonest genera among vascular plant introductions (Clement and Foster, 1994).

It is often the freshwater aquatic habitats which seem to fare worst when alien plants and animals arrive. Twenty-three known non-native aquatics are known to be established in Britain, the commonest, Canadian Waterweed *Elodea canadensis* first recorded in 1842, and the most recent, Floating Pennywort *Hydrocotyle ranunculoides*, first seen in 1990 and now found in ten 10 km squares (Farrell, 2001). In the ditches around Poole Harbour, Least Duckweed *Lemna minuta* and Water Fern *Azolla filiculoides* from the Americas, and Pygmyweed *Crassula helmsii* from New Zealand, have all been introduced.

As a result of the accidental introduction of the American Cord Grass *Spartina alterniflora* and hybridization with our native Cord Grass *Spartina maritima,* followed by the production of a fertile hybrid a few years later, large areas of Poole Harbour are covered by saltmarsh dominated by the fertile hybrid *Spartina anglica*, although this species has waned in the harbour over recent decades (Gray and Benham, 1990). This species is now seen as having some conservation benefit, as it provides some protection for saltmarshes against erosion, and to roosting waterbirds. *Spartina* is present, often as the only species, in over 50% of the high tide roosts used by wintering and passage waders in Poole Harbour (Morrison, 2004).

A marine non-native species directory published by the Joint Nature Conservation Committee (Eno *et al*., 1997) listed some 20 algal and 30 invertebrate species known to have become naturalized in British coastal waters, estuaries and lagoons. Difficulties associated with recording and monitoring species within the marine environment ensure that this is likely to be a considerable underestimation of the true number of marine non-natives that have become naturalized after introduction. Species are commonly introduced as ship fouling organisms, ballast water inhabitants, or associates of intentionally introduced shellfish.

Surveys undertaken in Poole Harbour over recent decades have revealed a substantial presence of marine seaweed and invertebrate invaders on shores and in low water channels (Dyrynda, 1987, 2003). Scattered shells of the North American Slipper Limpet *Crepidula fornicata* stranded on sandy shores around the harbour mouth originate from dense accumulations of limpet clusters that carpet low water channels in many parts of the harbour. This species is a serious pest of both natural and farmed oyster beds in Poole Harbour (Dyrynda, 1987) and many other locations in north-west Europe (Eno *et al.*, 1997). Copses of the Japanese Seaweed *Sargassum muticum* have been a feature of channel margins in the outer harbour during summer months since this species arrived in Poole *c.* 1977. The Korean Sea Squirt *Styela clava*, first recorded near Plymouth in 1953, now extensively colonizes mollusc shells and other substrates in Poole Harbour channels, whereas the introduced bivalve mollusc *Petricola pholadiformis* burrows in sub-tidal hard clay. The Australasian Barnacle *Elminius modestus* (introduced to the UK by Second World War shipping) is abundant on docks and piers across the harbour.

Marine invasions continue to the present day. In 1997, Poole Harbour provided the second Atlantic record for a Pacific bryozoan *Tricellaria inopinata* (Dyrynda *et al.*, 2000), and recent surveys (spring and summer 2003) have revealed two more significant records for Poole, pioneer colonies of Wakame *Undaria pinnatifida*, a large edible brown seaweed from South East Asia, and the extensive presence of the red seaweed *Agardhiella subulata* (Dyrynda, this volume, chapter 8). In most cases, these introductions have involved accidental transfers, usually via other British coastal locations. One exception, however, has involved hatchery-reared stock of the Manila Clam *Tapes phillipinarum*, farmed west of Brownsea Island since the late 1980s. Naturalization and colonization of sedimentary shores and shoals across the harbour followed during the 1990s, enabling the birth of a now well-established new commercial fishery.

Conclusion

The introduction of alien species is a complex issue. Effects on native habitats and species are often difficult to disentangle and many introductions appear to be neutral or benign. The problem in most cases is that we cannot predict when making a new introduction of an alien species, either deliberately or accidentally, whether it will be harmful to our native fauna and flora or not. When introduced species do prove to be harmful, eradication is, in most cases, not a realistic option, leaving acceptance, or resource-intensive, long-term control measures, as the only options.

Evidence from terrestrial studies shows that aliens are often a threat to native species, resulting in population declines, limiting distributions, and in extreme cases, causing extinctions (e.g. Clout and Veitch, 2002). It is less clear how introduced species impact on marine ecosystems and species. One estimate suggested that the main causes of extinction in marine organisms have been human exploitation and habitat destruction and that only a handful of species have been lost as a result of introductions (Dulvy *et al.*, 2003). However, marine extinctions and their causes are difficult to detect and could be underestimated. Experience suggests that in any consideration of introductions, the precautionary principle should always prevail.

References

Clement, E. J. and Foster, M. C. (1994) *Alien Plants of the British Isles*. London: Botanical Society of the British Isles.

Clout, M. N. and Veitch, C. R. (2002) Turning the tide of biological invasion: the potential for eradicating invasive species. pp: 1–3. In: *Turning the Tide: Eradication of Invasive Species. Proceedings of an International Conference on Eradication of Island Invasives.* Veitch, C. R. and Clout. M. N. (eds). Cambridge: World Conservation Union (IUCN).

Druce, G. C. (1928) *British Plant List*. Arbroath: Buncle.

Dulvy, N. K., Sadovy, Y. and Reynolds, J. D. (2003) Extinction vulnerability in marine populations. *Fish and Fisheries*, **4**: 25–64.

Dyrynda, P. E. J. (1987) *Poole Harbour Subtidal Survey. Baseline Assessment.* Report to the Nature Conservancy Council.

Dyrynda, P. E. J., Fairall, V. R., Occhipinti Ambrogi, A. and d'Hondt, J-L. (2000) The distribution, origins and taxonomy of *Tricellaria inopinata* d'Hondt & Occhipinti Ambrogi, 1985, an invasive bryozoan new to the Atlantic. *Journal of Natural History*, **34**: 1993–2006.

Dyrynda, P. E. J. (2003) *Marine Ecology of Poole Harbour*. http://www.swan.ac.uk/biodiv/poole

Eno, N. C., Clark, R. A. and Anderson, W. G. (1997) *Non-native Marine Species in British Waters: A Review and Directory.* Peterborough: Joint Nature Conservation Committee.

Farrell, L. (2001) Alien aquatic plants in Britain. In: *Alien Species – Friends or Foes? Journal of the Glasgow Natural History Society*, Suppl., **23**: 44–47.

Gray, A. J. and Benham, P. E. M. (1990.) *Spartina anglica – A Research Review.* London: HMSO and Institute of Terrestrial Ecology.

IUCN (2000) *Guidelines for the Prevention of Biodiversity Loss Due to Biological Invasion.* Gland: World Conservation Union.

Kent, D. H. (1992) *List of Vascular Plants of the British Isles*. London: Botanical Society of the British Isles.

Morrison, S. J. (2004) *Wader and Waterfowl Roost Survey in Poole Harbour 2002/03*. Wareham: Poole Harbour Study Group.

Rodwell, J. S. (ed.) (1991) *British Plant Communities*. Volume 1. *Woodlands and Scrub*. Cambridge: Cambridge University Press.

Young, B. (2000) Invasive species and other priorities for English Nature. In: *Exotic and Invasive Species*. Bradley, P. (ed.). *Proceedings of the Eleventh Conference of the IEEM, Winchester.*

The Ecology of Poole Harbour
John Humphreys and Vincent May (editors)

13. The Manila Clam in Poole Harbour

Antony Jensen[1], John Humphreys[2], Richard Caldow[3] and Chris Cesar[1]

[1]Southampton Oceanography Centre, University of Southampton, Hampshire SO14 3ZH

[2]University of Greenwich, Old Royal Naval College, Greenwich, London SE10 9LS

[3]Centre for Ecology and Hydrology, CEH Dorset, Winfrith Technology Centre, Dorchester, Dorset DT2 8ZD

The Manila Clam *Tapes philippinarum* was introduced to Poole Harbour in 1989 as a novel species for aquaculture. Contrary to expectations this species has become naturalized in the harbour, probably the northernmost location in Europe for this to occur. The chapter discusses possible reasons why this should have happened. The naturalized clam is now fished extensively in the western part of the harbour and preliminary data suggest that the pump-scoop method of fishing does not increase the degree of infaunal community disturbance (as measured by an ABC plot), but may have some effect on the sediment character.

Introduction

The Manila Clam *Tapes philippinarum* (Adams and Reeve 1850) is a sediment dwelling, bivalve mollusc of the family Veneridae. It is indigenous to western Pacific coastal seas (Goulletquer, 1997) and has over the years generated a good deal of taxonomic confusion. Consequently its scientific name has been volatile. It has at one time or another been located in seven different genera. Synonyms include *Tapes semidecussatus, Ruditapes philippinarum, Venerupis semidecussatus, Venerupis phillipinarum* and *Venerupis semidecussata* (Howson and Picton, 1997), among the 30 binomials used since 1791. Common names include the Littleneck Clam and the Japanese Palourde. The closest taxonomically related species native to UK waters is the Palourde or Carpet Shell *Tapes decussatus* (Canapa *et al.*, 1996), generically these two species are very different (Passamonti *et al.*, 1997).

Tapes philippinarum lives in fine sediments in the intertidal and upper sub-littoral zones. Relatively short inhalant and exhalent siphons are fused along much of their length – a feature, which in the UK distinguishes the Manila Clam from its indigenous taxonomic relative. It is commonly found 3–5 cm below the mud surface. The adult Manila Clam is tolerant of salinities down to 15 and the species distribution consequently extends into estuaries.

In the twentieth century, the distribution of the Manila Clam has been extended beyond its native coasts by human agency, occasionally inadvertently, but usually commercially motivated as the clam has considerable economic significance in terms of both fisheries and aquaculture (clam farming). Initial introductions were from Japan, the first in the 1920s to the Hawaiian Islands (Yap, 1977). In the 1930s, Manila Clams were accidentally introduced to the Pacific coast of North America (Quayle, 1964), mixed in with Pacific Oyster seed. They now range from California to Northern British Columbia (Magoon and Vining, 1981), being one of the principal commercially harvested clam species in Washington State. In the 1960s, they were imported to France for cultivation on both the Mediterranean and Atlantic coasts (Flassch and Leborgne, 1992). They have also been imported to Tahiti, the Adriatic coast of Italy, Germany and Spain. Since 1982, they have been cultivated in Ireland (Quishi Xie and Burnell, 1994).

The farming of clams commonly involves the laying of juvenile 'spat' under netting on an otherwise natural coastal or estuarine mudflat. Since spat can be grown and supplied commercially from hatcheries, successful clam farming can be achieved outside the range at which a natural breeding population can establish. Consequently, despite the presence of successful clam aquaculture enterprises in northern European waters, the capacity of the species to naturalize in the UK remains to be finally elucidated. Whilst Manila Clams have been found to mature and spawn in natural conditions in south-western Ireland (Qiushi Xie and Burnell, 1994), there does not appear to be a naturalized population in Ireland. In contrast Robert *et al.* (1993) reported on the growth, reproduction and gross biochemical composition of a naturalized Manila Clam population in the Bay of Arcachon, France. The first occurrence of a naturalized population in the UK has since been reported from Poole Harbour (Jensen *et al.*, 2004).

Naturalization in Poole Harbour

In 1980, the clam was introduced into the UK by the Ministry of Agriculture, Fisheries and Food (MAFF) in order to investigate the economic potential of the species in the UK. Whilst field trials demonstrated good potential, by 1992, the commercial production of the clam was still less than 50 tonnes per year. The slow growth of the industry was attributed in part to the activities of conservation groups including English Nature (the statutory nature conservation agency for England) who were concerned that the Manila Clam might naturalize and displace indigenous species (Utting, 1995). At that time, it was thought in MAFF that clams could not reproduce successfully in UK waters (Laing and Utting, 1994). Despite some evidence to this effect, conservation groups remained unconvinced and in this context MAFF initiated a programme of research at Conway to produce sterile triploid clams for farming.

Triploidy can be induced in bivalves by chemical or physical (heat or pressure) interventions during meiosis. The anticipated effect was sterility and improved meat production as a consequence of reduced resource-expensive gonad activity, as had earlier been reported for oysters. In the event, whilst triploid clams demonstrated reduced fecundity, sterilization was not reliably achieved.

In the context of this and other research programmes, MAFF conducted field trials at a number of locations including the Menai Straits in Wales, the estuary of the River Exe in Devon, and in Poole Harbour (Spencer *et al.*, 1996; Shpigel and Spencer 1996). The Manila Clam was introduced to Poole Harbour in 1988 by Othneil Shellfisheries following MAFF trials. By 1994, local fishers were exploiting the species, although it was not clear initially whether the clam had naturalized outside the aquaculture beds or if the spat fall was dependent on hatchery reared stock. It is now clear that both the stock on the aquaculture beds and in other areas of the harbour reproduce annually (Grisley, 2003).

The intertidal distribution of the clam in Poole Harbour was established in late 2002, during a systematic survey of 80 sampling stations along with additional sites sampled as part of a fisheries study (Figure 1). The figure also shows the sites of the original MAFF trials and the current Othneil Shellfisheries operation. It is clear from this pattern of distribution along with other studies, that a naturalized population of clams has achieved widespread distribution within the harbour, including Wareham Channel, Lytchett and Holes Bays, from Arne Bay south-east to Brands Bay, and on the northern coast at Parkstone Bay.

The Poole Harbour Environment

The existence of a naturalized population in Poole Harbour raises questions about the harbour environment and its compatibility with the needs of the Manila Clam: what is it about the harbour that has enabled this alien species to get established? Is the harbour unique in some respect or might this species colonize other parts of the UK coastline? In the light of the clams' natural distribution in sub-tropical and low boreal latitudes, water temperature is an obvious factor to examine.

Commonly limits to distribution are determined by environmental constraints on reproduction and recruitment. The physiological and other challenges of reproductive and early growth processes are more sensitive to environmental circumstances than is the physiology of mature individuals. It is well established that temperature is one of the most important environmental factors that influence reproductive activity in molluscs. Key processes in this respect are gametogenesis, spawning and larval growth. In the Manila Clam there is evidence that while gonad activity is possible down to 8 °C, the maturation of gametes does not occur below 12 °C. Spawning and larval growth requires temperatures of 14 °C while the growth of metamorphosed juveniles can occur at 10 °C. (Laing *et al.*, 1987; Laing and Utting, 1994).

Thus in UK waters, the processes of gamete production, spawning and larval growth are vulnerable. Since these are consecutive processes, a sustained period of higher temperatures is needed to achieve recruitment of viable Year 0 individuals to the population, and even if minimum limits are reached, the rate of these processes may be retarded. There is evidence that in North Wales, water temperature slows gametogenesis

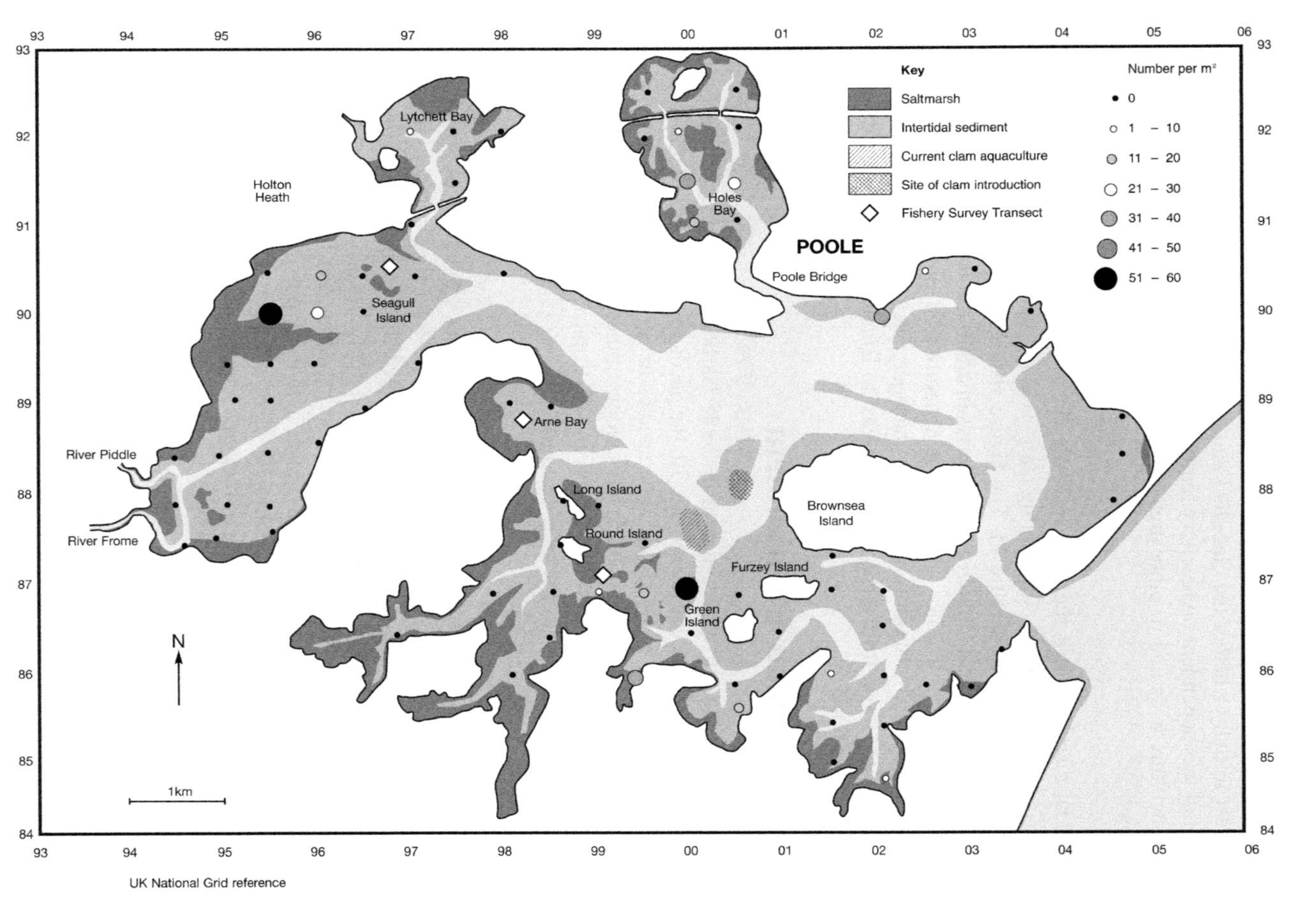

Figure 1 Distribution of the Manila Clam in 2002 survey.

with the consequence that spawning, even if it happens, is delayed to late summer, when larvae are unlikely to survive (Millican and Williams, 1985). The absence of reported naturalization at sites of field trials other than Poole Harbour, including the Exe estuary in Devon and in Ireland, suggests a similar explanation. Temperature data from areas of Poole Harbour (Grisley, 2003) with an intertidal population of Manila Clams combined with histological analysis of gonads shows that the temperature regime allowed ripe gametes to develop and spawning to start in June/July in 2000 and 2001. The successful survival and growth of the larvae is self evident from the success of the fishery and the presence of juvenile clams in benthic and fishery surveys within the harbour.

That the Manila Clam has naturalized in Poole Harbour may be related to features of the harbour that make it distinct from most other UK estuaries and sheltered coastal locations. The harbour is shallow, with small tidal range and, due to the narrow entrance, has a low flushing rate. This means that temperatures are not moderated much by exchange of water with the open sea. This along with its south coast location makes the harbour in summer probably amongst the warmest substantial bodies of seawater around the UK.

If we are correct in our hypothesis that the ability of the clam to naturalize in Poole Harbour is linked to the distinctive nature of the harbour environment, then we would not expect the species to proliferate extensively on the UK coast. It remains possible, however, that certain somewhat similar UK harbour environments may in due course be colonized.

Clam fishing in the harbour

The presence of a naturalized clam population in the harbour attracted the attention of Poole fishermen and by 1994, a licensed (by the Southern Sea Fishery Committee (SSFC)) fishery had become established. The fishery for clams is focused on the intertidal mudflats of the western part of the harbour and (following a SSFC restriction on hand picking in Poole Harbour to avoid disturbance to feeding birds at low tide after concerns expressed by English Nature and RSPB) is undertaken from boats working during daylight high tides during late October, November, December and early January. Clam harvesting was initially achieved by the use of a 'clam scoop' – a toothed mesh basket open at one end with a long (*c.* 2.5 m) handle (Figure 2). The scoop is pushed into the sediment and pulled along by the forward motion of the boat. Sediment is removed from the scoop by an up and down motion on the handle as the scoop passes through the sediment. This is hard physical work and naturally fishers looked at ways to improve catch efficiency with less physical effort. Over a number of years the current 'pump-scoop' was developed (Figure 3), the major difference being that a spray of water (from an engine-powered pump) is placed at the mouth of the scoop which washes sediment out of the basket, so removing the need for the manual up and down motion on the scoop handle. This is a system almost unique to Poole Harbour and should not be confused with suction or hydraulic dredging techniques which differ in that they both fluidize the sediment by inserting water into the sediment immediately in front of the suction device or dredge. A variation of this technique was developed to work within the harbour with cockle dredges.

Figure 2 **A Poole Harbour clam scoop. This scoop has been modified with a finer internal mesh to retain smaller clams than taken commercially to facilitate scientific survey work.**

Figure 3 **A Poole Harbour clam pump-scoop. Note how the water jets enter the front of the scoop and wash the sediment out of the basket.**

Potential impact of pump-scoop fishing in the harbour

Concerns were expressed by environmental groups, predominantly English Nature and RSPB, that the pump-scoop fishing technique may be: (i) disturbing roosting birds by creating significant noise; and (ii) impacting on the amount and type of intertidal infauna (mostly marine worms and bivalves) available for overwintering birds. Whilst (i) could be dealt with by noise abatement orders, (ii) required some further research to quantify the impact of the pump-scoop fishing technique in the harbour. The impact of cockle fishing with a pump-dredge is considered elsewhere in this volume (see chapter 17), some preliminary results relating to the clam pump-scoop fishery in 2001/02 are described below (Cesar, 2003).

Three replicate 0.06 m^2 sediment samples were taken from 13 sites in Poole Harbour before and after the clam fishing season in 2002/03 (and a 14 sample site survey has been completed for 2003/04). The 2002/03 samples were sieved on a 1 mm sieve and preserved in 4% formalin. Separate samples were taken to establish sediment granulometry. Data from four sites have been analysed to date (February 2004) and data from a single site (Seagull Island) that experienced 'high' fishing pressure (as designated by district fishery officers) is shown here as an example. Preserved infaunal samples were sorted, identified and counted and these data were analysed using the multivariate analysis package PRIMER (Plymouth Routines In Multivariate Ecological Research) (Clarke and Warwick, 1994). To establish dry weight data, animals were grouped by taxonomic family and dried at 60 °C for 24 hours. Biomass and abundance were then compared using the Dominance plot program in PRIMER (Clarke and Warwick, 1994), which generates abundance-biomass comparison plots and allows a qualitative (W statistic) and quantitative estimation of biological community disturbance (if abundance plots out above biomass the community is dominated by short-lived species and is likely to be disturbed, if biomass plots out above abundance the community is dominated by long-lived species and is unlikely to be in a disturbed state). Sediments were wet sieved to separate fine sediments (<63 mm) from coarse (>63 mm). Coarse sediments were then dry sieved and fine sediments analysed using a calibrated LS130 Coulter counter. It was expected that if sediment granulometry altered during the fishing season, the fine sediments would show most change.

Abundance-Biomass Comparison (ABC) plots for Seagull Island (Figure 4) show the infaunal community to have a qualitatively similar level of disturbance before and after the pump-scoop fishing season. The W statistic values generated by this method (18 species and average W value of 0.325 before fishing, 15 species and average W value of 0.297 after fishing at Seagull Island) for all four sites that were sampled before and after fishing were tested statistically by Analysis of Variance (ANOVA) and found not to be statistically different ($F = 0.348$, $a = 0.05$) (Cesar, 2003).

Fine sediment (<63 mm) granulometry (Figure 5) also shows some quantitative change in the period between sampling at Seagull Island but when sediment samples from all four sites were subjected to a χ^2 test, no significant difference was seen at the 5%

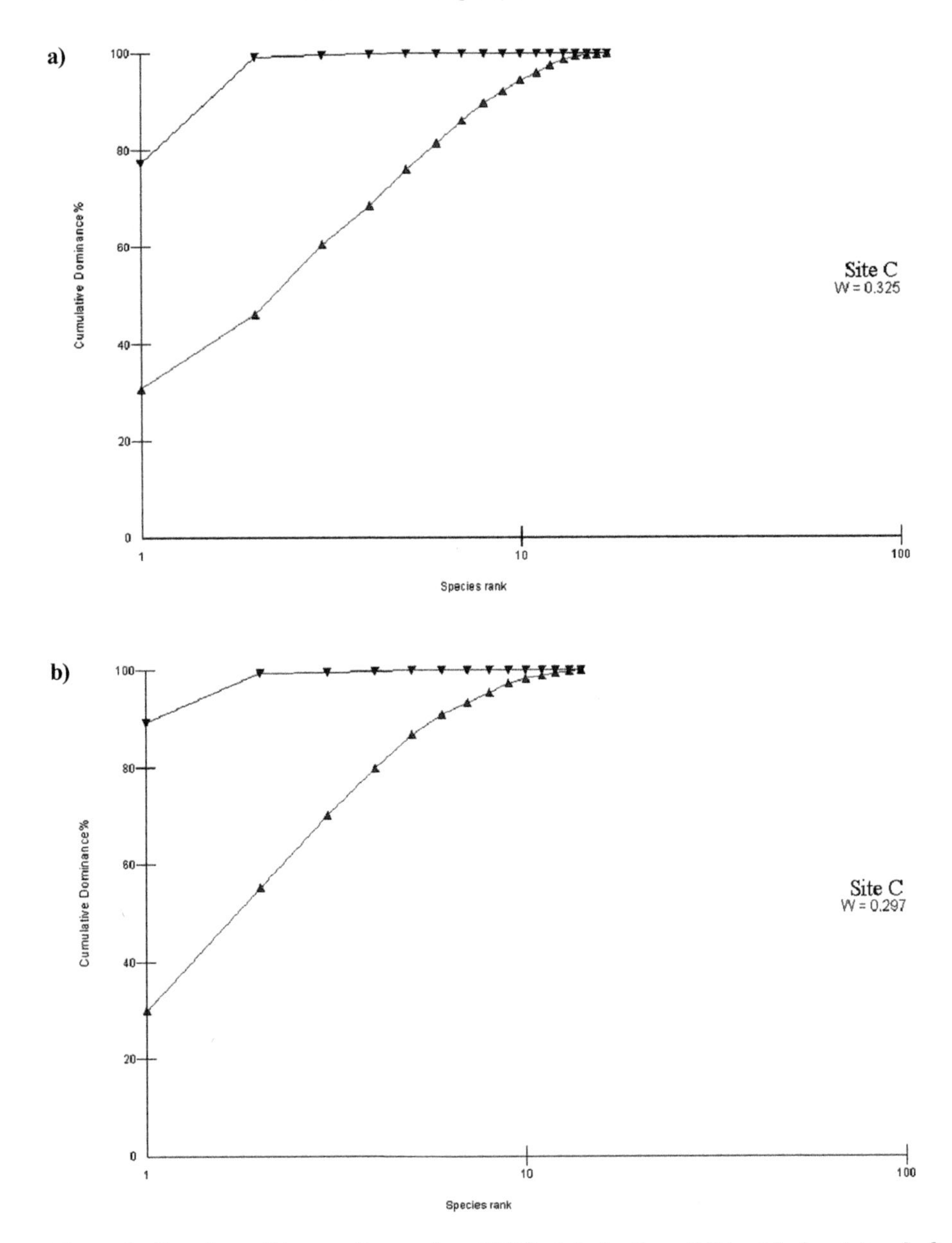

Figure 4 Abundance-Biomass Comparison (ABC) plots for Seagull Island before (a) and after (b) the 2001/02 fishing season: ▲ = abundance, ▼ = biomass (Cesar, 2003).

significance level between before and after fishing season granulometry (Cesar, 2003). Data such as the ABC plots and fine sediment comparisons allow some assessment to be made of the changes happening in the harbour over time. It would seem that some qualitative changes had occurred to sediments at Seagull Island but these were much less

Pre-fishing sample

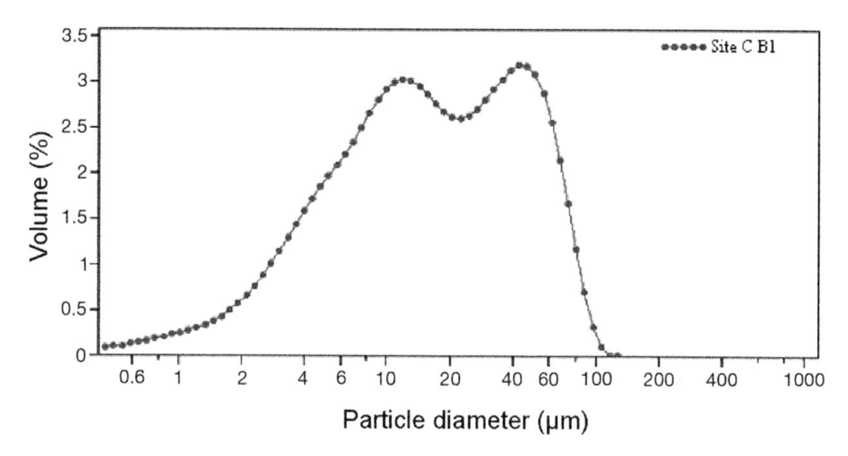

Post-fishing sample

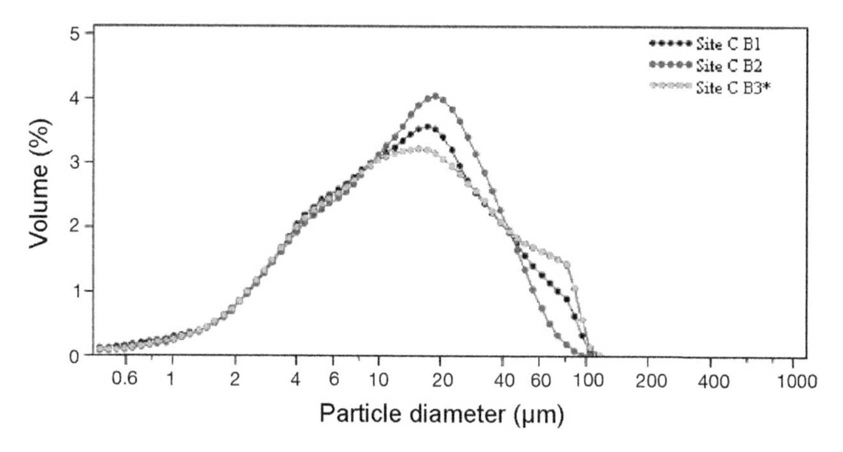

Figure 5 Fine sediment granulometry before and after the 2001/02 fishing season at Seagull Island. Note replicate samples were not taken for this site pre-season.

at the other three sites surveyed. Conclusions drawn from these preliminary data must be tentative but it would seem that there was no significant additional disturbance to the infaunal community, as measured by ABC plots, before and after the 2002/03 fishery season in the four sites sampled. However, whilst not statistically significant, there was a suggestion at one site in particular (Seagull Island), that some changes in the sediment granulometry had occurred between the dates sampled. The remainder of the samples for 2002/03 and 2003/04 now need to be analysed to provide a clearer assessment of what alterations, if any, are occurring in the harbour areas subject to pump-scoop fishing activity.

Acknowledgements

The authors would like to thank English Nature, Southern Sea Fisheries Committee, the RNLI and SOC for support whilst undertaking the work that contributed to this chapter.

References

Canapa, A., Marota, I., Rollo, F. and Rollo, E. (1996) Phylogenetic analysis of Veneridae (Bivalvia): Comparison of molecular and palaeontological data. *Journal of Molecular Evolution,* **43**: 517–522.

Cesar, C. P. (2003) *The Impact of Clam Fishing Techniques on the Infauna of Poole Harbour.* Dissertation submitted in partial fulfilment of the requirements for the degree of MSc (Oceanography), School of Ocean and Earth Science, University of Southampton.

Clarke K. R. and Warwick, R. M. (1994) *Change in Marine Communities: An Approach to Statistical Analysis and Interpretation*. Plymouth: Plymouth Marine Laboratories.

Flassch, J. P. and Leborgne, Y. (1992) Introduction in Europe, from 1972 to 1980, of the Japanese clam (*Tapes philippinarum*) and the effects on aquaculture production and natural settlement. *ICES Marine Symposium,* **194**: 92–96.

Grisley, C. (2003) The Ecology and Fishery of Tapes philippinarum (Adams and Reeve, 1850) in Poole Harbour, UK. A thesis presented for the degree of Doctor of Philosophy, School of Ocean and Earth Science, University of Southampton.

Goulletquer, P. (1997) *A Bibliography of the Clam (Tapes philippinarum*). La Tremblade: IFREMER.

Howson, C. M. and Picton, B. E. (eds) (1997) *The Species Directory of the Marine Fauna and Flora of the British Isles and Surrounding Seas*. Ulster Museum publication No. 276. Belfast: Ulster Museum/Ross on Wye: Marine Conservation Society.

Jensen, A. C., Humphreys, J., Grisley, C., Caldow, R. W. G. and Dyrynda, P. (2004) Naturalisation of the clam *Tapes philippinarum* an alien species and the establishment of a clam fishery within Poole Harbour, Dorset. *Journal of the Marine Biological Association of UK.* **84**: 1069–1073.

Laing, I. S. D. and Utting, S. D. (1994) The physiology and biochemistry of diploid and triploid clams (*Tapes philippinarum*) larvae and juveniles. *Journal of Experimental Marine Biology and Ecology*, **184**: 159–169.

Laing, I. S. D., Utting, S. D. and Kilada, R. W. S. (1987) Interactive effect of diet and temperature on the growth of juvenile clams. Journal *of Experimental Marine Biology and Ecology*, **113**: 23–28.

Magoon, C. and Vining, R. (1981) *Introduction to Shellfish Aquaculture in the Puget Sound Region, WDNR, Olympia, WA (USA). 79pp*

Millican, P. F. and Williams, D. R. (1985) *The Seasonal Variation in the Levels of Meat Content, Lipid and Carbohydrate in Mercenaria mercenaria and Tapes semiducussatus Grown in Fertilised and Unfertilised Water*. CM 2985/K.51. Copenhagen: International Council for the Exploration of the Sea.

Passamonti, M., Mantovani, B. and Scali, V. (1997) Allozymic characterization and genetic relationships among four species of Tapetinae (Bivalvia, Veneridae). *Italian Journal of Zoology,* **64**: 117–124.

Qiushi, Xie and Burnell, G. M. (1994) A comparative study of the gametogenic cycles of the clams *Tapes philippinarum* and *Tapes ducussatus* on the south coast of Ireland. *Journal of Shellfish Research*, **13** (2): 467–472.

Quayle, D. B. (1964) Distribution of introduced marine molluscs in British Columbia waters. *Journal of Fish Research Board, Canada,* **21** (5): 1155–1181.

Robert, R., Trut, G. and Laborde, J. L. (1993) Growth, reproduction and gross biochemical composition of the clam *Ruditapes philippinarum* in the Bay of Arcachon, France. *Marine Biology*: 291–299.

Shpigel, M. and Spencer, B. (1996) Performance of diploid and triploid Clams *Tapes philippinarum* at various levels of tidal exposure in the UK and in water from fish ponds at Eilat, Israel. *Aquaculture*, **141**: 259–171.

Spencer, B. E., Kaiser, M. J. and Edwards, D. B. (1996) The effect of clam cultivation on an intertidal benthic community: the early cultivation phase. *Aquaculture Research*, **27**: 261–276.

Utting, S. D. (1995) Triploidy in the clam *Tapes philippinarum. Workshop on the Environmental Impacts of Aquaculture Using Organisms Derived Through Modern Biotechnology*. Paris: OECD.

Yap, W. G. (1977) Population biology of the Japanese littleneck clam, *Tapes philippinarum*, in Kaneohe Bay, Oahu, Hawaiian Islands. *Pacific Science,* **31** (3): 223–245.

The Ecology of Poole Harbour
John Humphreys and Vincent May (editors)

14. Ecological Impacts of Sika Deer on Poole Harbour Saltmarshes

Anita Diaz, Eunice Pinn and Justine Hannaford

School of Conservation Sciences, Bournemouth University, Fern Barrow, Poole, Dorset BH12 5BB

This study investigates the effect of an introduced species, Sika Deer *Cervus nippon* on saltmarsh plant and infaunal communities. Epidermal fragment analysis was used to identify the plant species eaten by Sika Deer. Vegetation communities in deer exclosures and openly grazed areas were monitored over 4 years to investigate the effect of grazing on plant community composition and structure. The infaunal communities were assessed by extracting sediment cores. Deer were found to graze preferentially on *Spartina anglica* as intensive grazing led to swards dominated by *Salicornia ramosissima* even in upper marsh areas. Highest plant diversity was related to intermediate levels of grazing. Overall, higher levels of grazing led to higher abundance of three species of infauna detected in this study: *Hydrobia ulvae*, *Gammarus* sp. and *Nereis diversicolor*. Detailed examination revealed that the high abundance of *H. ulvae* was related to small quantities of above ground vegetation volume and that the abundance of *Gammarus* was related to small quantities of below ground vegetation biomass. The possible direct and indirect effects of Sika Deer grazing on bird populations are discussed.

Introduction

Sika Deer *Cervus nippon* are native to Japan and East Asia and were first brought to British deer parks and private collections during the 1800s (Putman, 2000; Whitehead, 1964). The first introductions of Sika Deer to Dorset appear to have occurred in 1896 at Brownsea Island and during the early twentieth century, when they were brought to Hyde House, north of Wareham (Mitchell-Jones and Kirby, 1997; Whitehead, 1964). Animals brought to Brownsea Island are reported to have escaped by swimming ashore during their first night on the island (Whitehead, 1964) and animals escaped from Hyde House into the surrounding countryside throughout the early twentieth century and, in particular, during the Second World War (Horwood and Masters, 1981). It is not certain how many deer were originally brought to Purbeck nor where they were brought from, but it seems most likely that they originated from stock bred at Powerscourt in County Wicklow (Horwood and Masters, 1981). Sika Deer were brought to Powerscourt directly from Japan in 1861 and it is likely that these deer bred with other species of *Cervus* held at Powerscourt during the late 1800s, including Red Deer *Cervus elephus* and Sambar *C. unicolor* (Ratcliffe, 1987). Consequently the Sika Deer now feral in Purbeck are likely to exhibit a degree of hybridization (Ratcliffe, 1987).

The Isle of Purbeck currently has the largest group of feral Sika Deer in England (Putman, 2000) and deer from Purbeck are spreading further round Poole Harbour. The total number of Sika Deer in Purbeck was estimated as several hundred over a decade ago (Mann and Putman, 1989), approximately 2000 animals a few years ago (Putman, 2000) and perhaps as many as 3000 at present (Hann, pers. comm.). Poole Harbour and its surrounding countryside contain a rich mosaic of internationally important wildlife habitats and so it is crucial to be able to determine the impact of Sika Deer on these habitats and to have an effective deer management strategy.

One of the most important habitats in the Poole Basin is the intertidal flats and saltmarshes. The international importance of these in providing feeding and roosting grounds for large numbers of wintering wildfowl and waders have contributed to the harbour being designated a Special Protection Area (SPA) under the European Birds Directive and a RAMSAR Site. The saltmarshes are also important during the spring and summer as breeding sites for waders, gulls and terns. The saltmarshes in Poole Harbour are composed largely of *Spartina anglica*, a species of grass that evolved following an initial hybridization event between a native species S. *maritima* and an American introduction *S. alternifolia*. As commonly occurs in evolution of a new, fertile species by chromosome doubling (polyploidy), *S. anglica* was more vigorous than its parental species and so *S. anglica* is now the most abundant species of *Spartina* in Poole Harbour. It is an interesting coincidence that at the turn of the last century, two species arrived in Poole Harbour with enormous potential for impact on ecosystem function; a competitive plant *S. anglica* and a large herbivore, *C. nippon.* One hundred years on, both species are abundant around the harbour and this raises challenging questions on the future for nature conservation grazing management of the Poole Harbour saltmarshes.

Arne RSPB reserve is located on the western edge of Poole Harbour and covers approximately 535 ha of saltmarsh, heathland, woodland and farmland. Arne appears to have a fluctuating but large population of Sika Deer (numbers for 2003 estimated at 500–700; Gartshore, pers. comm.) and deer can commonly be seen grazing on the saltmarshes. The saltmarshes provide winter feeding and roosting for large numbers of waders and wildfowl and breeding grounds for Redshank *Tringa totanus* (Price, 1997). The aim of the study reported here is to investigate the effect of Sika Deer on aspects of the saltmarsh ecosystem that may impact on its ability to support wildfowl populations. Four specific questions are asked:

(i) What saltmarsh plants do Sika Deer eat?
(ii) What effect do deer have on the plant species composition of the saltmarshes?
(iii) What effect do deer have on the vegetation volume of the saltmarshes?
(iv) What effect do deer have on the infauna abundance of the saltmarshes?

Methods

Saltmarsh plant species eaten by Sika Deer

Plants eaten by Sika Deer were identified by examination of epidermal fragments remaining in faecal pellets. Forty fresh faecal pellets were collected from the saltmarshes alongside Crichton's Heath North and South (Figure 1) in each of four weeks during July 2000. Pellets were collected at random with the restriction that only one pellet was taken from each pile of pellets to avoid pseudo-replication. Pellets were frozen on the day of collection to prevent decomposition of the epidermal fragments.

The epidermal fragments in each pellet were analysed by comparison against plant species held in an epidermal library at Bournemouth University. Species were scored as either present/absent in each of the 160 pellets to obtain a frequency of occurrence. The abundance of each species within a pellet was not measured as this can be greatly influenced by the relative digestibility of different plant species (Mitchell *et al.*, 1977). Pellets were prepared for analysis as described in Mack (2001).

A calculation was made of mean frequency of occurrence across weeks in July of each saltmarsh plant species in the epidermal fragments and this was related to the availability

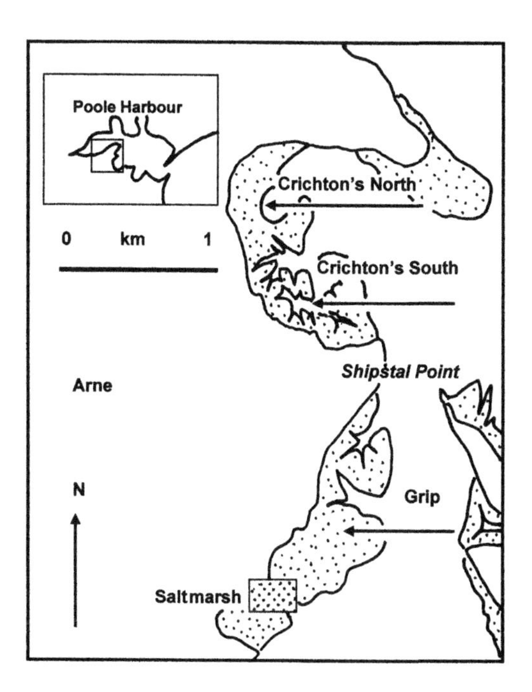

Figure 1 Locations of the three areas of saltmarsh studied.

of each species on the saltmarsh. The abundance of each plant species on these areas of saltmarsh was assessed by recording the vegetation in 20 randomly located 2 m x 2 m quadrats.

Effect of deer on the plant species composition of the saltmarshes

In May 1999, 10 deer exclosures were set up in random positions on heavily grazed areas of saltmarsh at Arne (five on the saltmarsh alongside Crichton's Heath North and five on the saltmarsh alongside Crichton's Heath South (Figure 1). Each exclosure measured 2.5 m x 2.5 m and was constructed using 2 m wooden stakes inserted to a depth of 1 m at each corner. Chicken wire was then used to enclose and roof the plot. To minimize edge effects, only the central 2 m x 2 m area of each exclosure was used for data collection. Ten unexclosed plots were also set up at random on heavily deer grazed areas of saltmarsh. The vegetation community on each of the 20 plots was assessed in early July of 1999, 2000, 2001, 2002 and 2003 by recording the percentage cover of each plant species present.

During the summer of 2002, a survey of Arne saltmarsh was undertaken to assess the general extent of deer grazing at Arne. From this survey, areas that were lightly or ungrazed between 1999 and 2002 were identified, and 20 lightly or ungrazed 2 m x 2 m plots (hereafter referred to as ungrazed) were set up and their plant community composition measured as above. Five of these plots were located at Crichton's Heath North and five at Crichton's Heath South.

Detrended Correspondence Analysis was used to ordinate the differences in plant communities. Similarities and relative dissimilarities between plant assemblages were calculated using Similarity Percentages (SIMPER) and cluster analysis within the package PRIMER (Plymouth Routines In Multivariate Ecological Research; Clark, 1993). The Shannon Index was used to measure differences in plant community diversity.

Effect of deer on the vegetation volume of the saltmarshes

In July 2002, above ground vegetation volume was assessed in each of the above 30 plots by visually recording the percentage occupancy of slices of the plot cuboid at 10 cm height intervals. Above ground volume was also assessed in 20 new 2 m x 2 m plots located on Grip Heath saltmarsh. Ten of these plots were located at random in heavily grazed areas of Grip Heath and 10 were located at random in lightly or ungrazed areas. To assess the possible impact of deer grazing below ground, root biomass was also investigated in each of the 50 plots. A 20 cm diameter augur drill was used to obtain three core samples to a depth of 10 cm. These samples were sieved using a 0.5 mm mesh to retain the root biomass. Where necessary, i.e. when the core contained large amounts of root biomass, cores were sub-sampled. The root biomass obtained was washed clean and transferred to an oven at 70 °C to dry for 48 hours, after which the samples were weighed. All data were analysed using a Kruskal-Wallis test comparing grazed, ungrazed and fenced sites. The relationship between above ground vegetation cover and below ground root biomass was assessed using a Spearman's Rank Correlation.

Effect of deer on the infauna abundance of the saltmarshes

The abundance of the macro-infauna of the saltmarsh was also assessed using the augur drill. The cores collected from the 50 plots described above were sieved through a 0.5 mm sieve and retained invertebrate fauna were identified to species level. These data were analysed using Analysis of Similarities (ANOSIM) in PRIMER.

Results

Saltmarsh plant species eaten by Sika Deer

Table 1 shows the mean percentage frequency of occurrence of each saltmarsh species in the deer pellets examined and the relationship of this to the abundance of each species on the saltmarsh. The most abundant species on the marsh were *Spartina anglica*, *Puccinella maritima* and *Salicornia ramosissima*. The species most frequently recorded in the pellets was *S. anglica*. The occurrence in the diet as a ratio of occurrence on the marsh is larger for *S. anglica* than for *P. maritima* while *Salicornia ramosissima* was seldom found in the faecal pellets despite its ready availability on the marsh. The amount of *Atriplex portulacoides* found in pellets as a ratio of that available on the marsh was intermediate to that for the above species.

Table 1 Frequency of occurrence of saltmarsh plant species on the marsh (mean ± SE from 20 2 m x 2 m quadrats) and in Sika Deer faecal pellets*

Species	Frequency of occurrence				Ratio of mean occurrence pellet/marsh
	Species on marsh		Species in pellets		
	mean	SE	mean	SE	
Aster tripolium	0.1	0.1	0.0	0.0	0.00
Atriplex portulacoides	11.6	3.2	16.875	8.4375	1.46
Limonium spp.	0.4	0.3	0.0	0.0	0.00
Plantago coronopus	0.1	0.1	0.0	0.0	0.00
Plantago maritima	0.2	0.1	0.0	0.0	0.00
Puccinella maritima	30.8	5.4	52.5	8.9	1.71
Salicornia ramosissima	19.4	5.9	7.5	1.5	0.39
Spartina anglica	40.6	8.5	89.4	2.6	2.20
Suaeda maritima	0.4	0.3	0.0	0.0	0.00
Triglochin maritima	0.3	0.3	0.0	0.0	0.00

* From a total of 160 pellets collected on four occasions.

Table 2 Analysis (SIMPER) of the factors accounting for the differences in the floral community of fenced and unfenced plots in heavily grazed areas in 2002

Variable	Percentage abundance		Cumulative percentage
	grazed	fenced	
Spartina anglica	10.85	85.5	40.57
Bare mud	38.25	5.0	59.56
Salicornia ramosissima	34.65	4.0	76.33
Puccinella maritima	26.4	18.5	88.92
Atriplex portulacoides	2.25	17.0	97.44

Effect of deer on the plant species composition of the saltmarshes

The areas fenced for 4 years showed a gradual shift in vegetation communities over time from *Salicornia ramosissima* dominated swards to *Spartina anglica* dominated swards (Figure 2). SIMPER analysis showed that almost half of the difference between the fenced and grazed swards was accounted for by the much lower abundance of *S. anglica* in the grazed swards (Table 2). Comparison of fenced, grazed and ungrazed plots in 2002 showed that, 3 years after being fenced, plots in previously heavily grazed areas had developed a vegetation that was not significantly different from the ungrazed plots (Figure 3). Plant species diversity was found to be greatest in plots where the level of vegetation volume (hence grazing pressure) was intermediate (Figure 4).

Effect of deer on the vegetation volume of the saltmarshes

Across all sites, above ground vegetation volume was found to be significantly lower in the grazed plots than in the ungrazed or fenced plots (Figure 5, $K = 37.60$, $P < 0.001$). Mean root biomass was found to be significantly lower in the grazed plots and greater in the ungrazed and fenced plots (Figure 6, $K = 23.69$, $P < 0.001$). As above ground vegetation increased so did the below ground root biomass (Figure 7) ($r = 0.68$, $P < 0.001$).

Effect of deer on the infauna abundance of the saltmarshes

Only three infaunal species were observed in the current study: a snail *Hydrobia ulvae*, a crustacean *Gammarus* sp. and the ragworm *Nereis diversicolor*. All species were found to be most abundant in the grazed plots and least abundant in the fenced plots (Figure 8). ANOSIM found the differences in invertebrate abundance in relation to grazing regime to be significant (Global $R = 0.126$, $P < 0.01$). Reasons for this relationship were explored by comparing invertebrate abundance to abundance of above and below ground vegetation using partial correlation analysis to hold one factor constant whilst the relationship of invertebrates with the other factor is quantified. This revealed that *Hydrobia ulvae* was

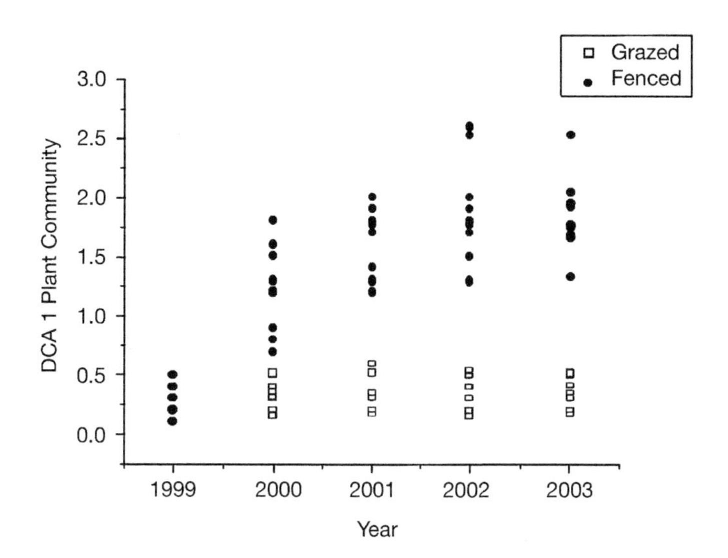

Figure 2 **Changes in plant community composition (DCA 1) over time (1999–2003) for fenced and unfenced plots in heavily grazed areas. DCA 1 represents a gradient of communities where communities with low DCA values are dominated by *Salicornia ramosissima* and those with high values for DCA 1 are dominated by *Spartina anglica.***

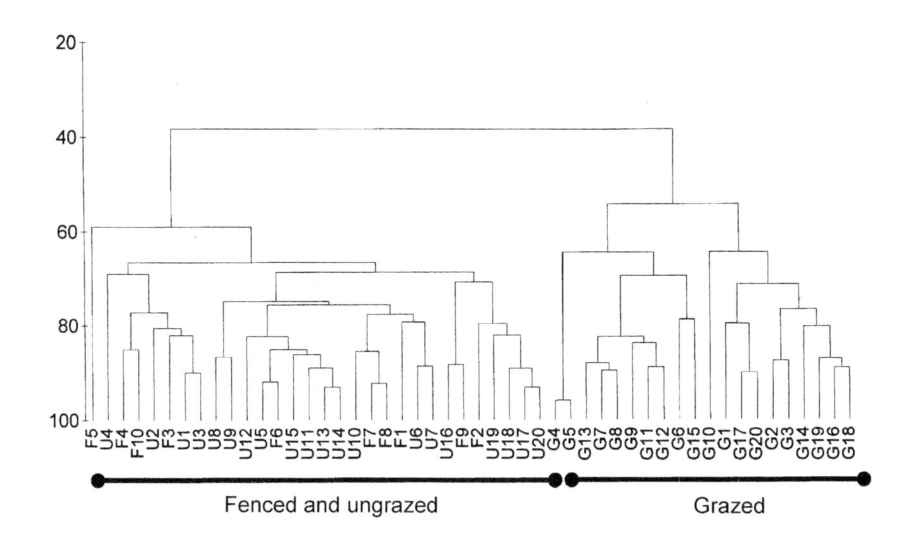

Figure 3 **Degree of similarity of plant community composition between fenced plots (F), unfenced plots in grazed areas (G) and unfenced plots in ungrazed areas (U) in 2002. Plots fenced for 4 years have a vegetation similar to those of plots in ungrazed areas.**

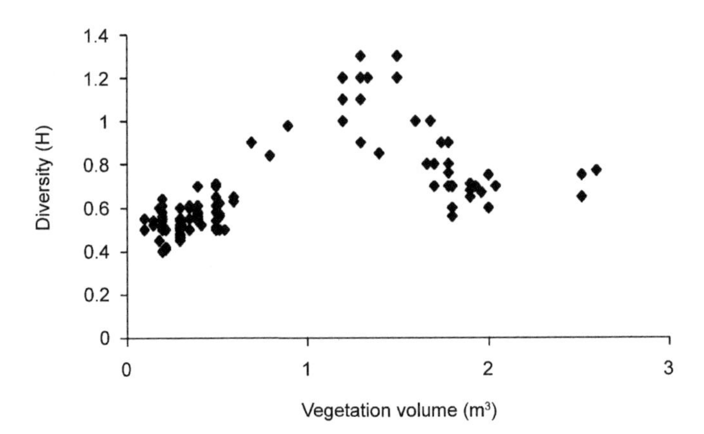

Figure 4 Relationship between grazing level (measured in terms of remaining vegetation above ground volume) and plant species diversity (measured by Shannon Index (H)).

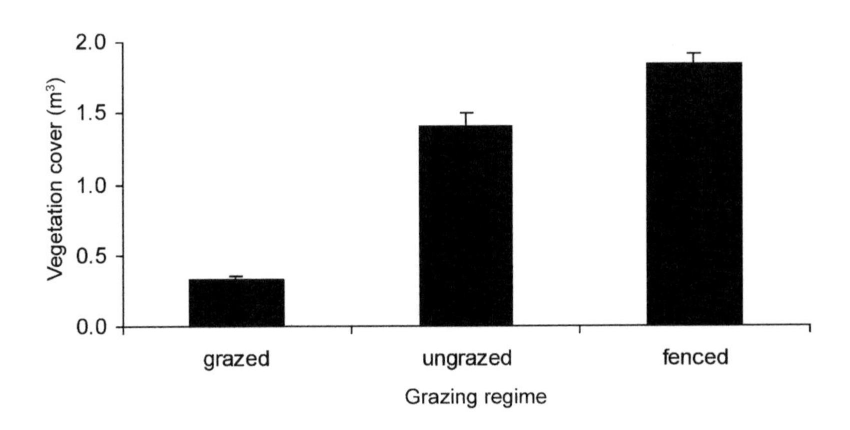

Figure 5 Above ground vegetation volume (mean ± SE) in grazed, ungrazed and fenced plots in 2002.

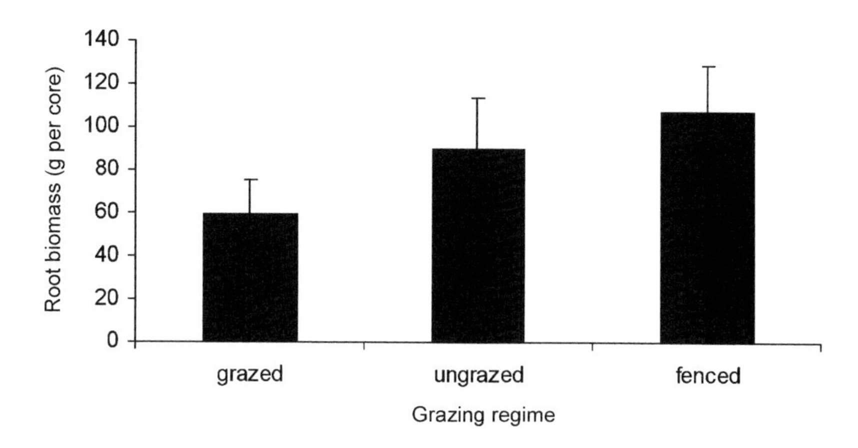

Figure 6 Below ground vegetation (roots) volume (mean ± SE) in grazed, ungrazed and fenced plots in 2002.

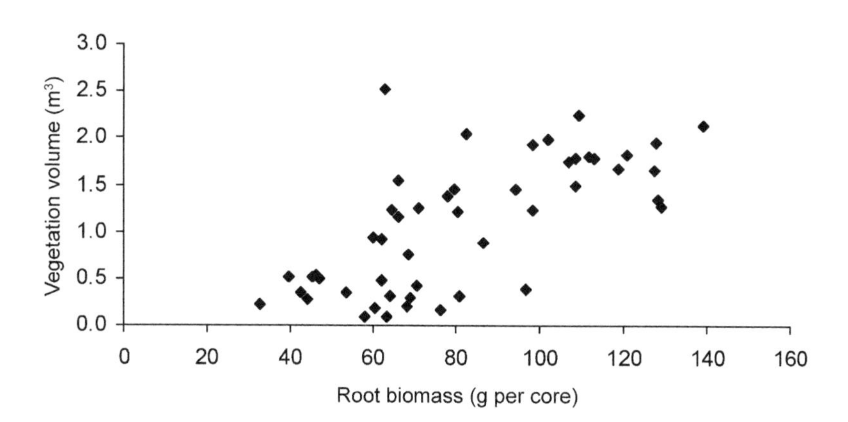

Figure 7 Relationship between above ground and below ground vegetation volume in 2002.

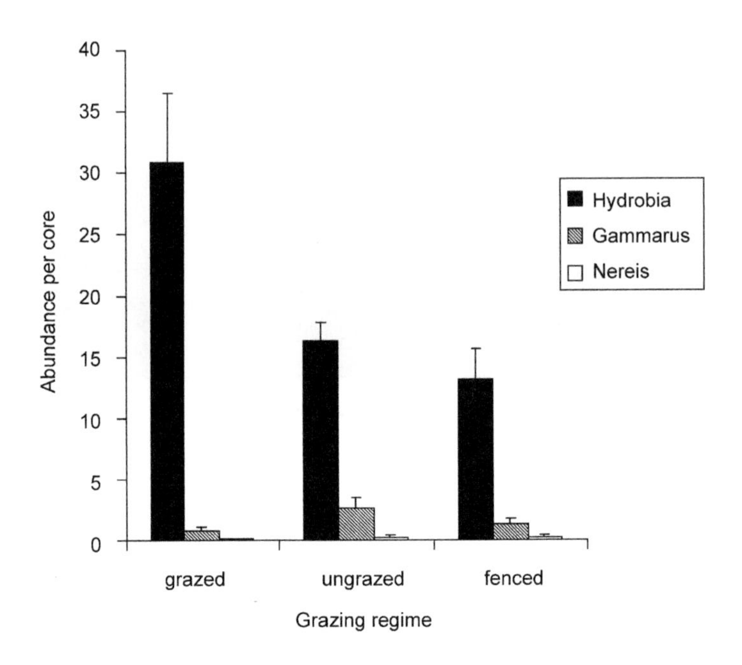

Figure 8 Abundance of infauna (mean ± SE) in grazed, ungrazed and fenced plots.

most abundant where there was least above ground biomass, irrespective of the amount of below ground biomass (Table 3). By contrast, *Gammarus* sp. numbers appeared unaffected by the amount of above ground biomass but abundance was greatest in areas of relatively low below ground biomass (Table 3). *Nereis* abundance showed no relationship with either above or below ground biomass. However, this result may be caused by low statistical power due to the small numbers of *Nereis* in our samples.

Discussion

Sika Deer were found to have a significant impact on the saltmarsh flora and fauna at Arne. Where grazing was severe, there was an almost total loss of vegetation cover. Species that respond best to grazing are those such as *Salicornia ramosissima* that are quickly able to colonize the gaps created (Bertness, 1991; Pehrsson, 1988) and appear to be unpalatable to grazers. The creation of bare patches in saltmarshes can lead to the development of hypersaline conditions due to increased surface evaporation in the absence of vegetation cover (Bertness, 1991). This may have contributed to the development of lower saltmarsh communities, able to tolerate high levels of salinity during the summer, dominated by *Salicornia ramosissima*, in the intensively grazed areas of the higher marsh positions at Arne.

Table 3 Partial correlation coefficients showing relationship between infauna abundance and (i) vegetation volume and (ii) root biomass

	Partial correlations	
	For above ground vegetation volume	**For below ground root biomass**
Hydrobia	r = -0.3567 P = 0.012	r = 0.0700 P = 0.633
Gammarus	r = 0.2026 P = 0.045	r = -0.2876 P = 0.045
Nereis	r = 0.0116 P = 0.938	r = 0.1167 P = 0.425

The main saltmarsh species eaten by Sika Deer was found to be *Spartina anglica* and this was consumed in preference to *Puccinella maritima*. The tolerance of *Puccinella maritima* to grazing has been reported in other studies carried out in several countries (Esselink *et al.,* 2000; Jensen, 1985; Bakker, 1978; Cadwalladr *et al.,* 1972). Consequently there is strong evidence that further intensification of grazing in the Poole Harbour saltmarshes may result in a change in the dominant plant species from *Spartina anglica* to *Puccinella maritima*. Similarly, results from this study and inference from the findings of other studies indicate that a reduction in grazing would be expected to lead to a further domination of the marsh by *Spartina anglica* except in areas around the creeks and low saltmarsh where the dominant species would likely be *Atriplex portulacoides* (Andresen *et al.*, 1990; Jensen, 1985; Bakker and Ruyter, 1981).

In this study, grazed plots regained the species composition typical of ungrazed plots within 4 years. This rapid rate of recovery is encouraging at face value, but is likely to be, in part at least, fuelled by the ready availability of propagules from nearby ungrazed areas. It is clear also that a total exclusion of grazers is not beneficial in terms of plant diversity, as this was highest in areas with intermediate levels of grazing. The actual grazing pressure exerted in such areas (deer $ha^{-1}day^{-1}$) is difficult to ascertain accurately but is currently being investigated. However, the general finding agrees with other studies that show that diversity is maximized by intermediate levels of grazing (Bouchard *et al.,* 2003; de Leeuw and Bakker, 1986; Bakker, 1985; Jensen, 1985). A high alpha and beta diversity of plant species may also lead to higher diversity in other trophic levels. For example, high densities of Redshank are closely correlated with a diverse vegetation community and structure (Norris *et al.,* 1997) and the highest densities of Redshank have been associated with moderately grazed marshes (Allport *et* *et al.,* 1998).

A reduction in above ground vegetation cover as a result of grazing is perhaps an obvious finding, and one that has been reported elsewhere (e.g. Bakker, 1978; Morris and Jensen,

1998; Esselink *et al.*, 2000; Seliskar, 2003). The current study also found that grazing affected below ground biomass. Although this reduced allocation of tissue to roots concomitant with loss of above ground biomass is well documented (Morris and Jensen, 1998; Belsky, 1987), the relationship between changes in above ground primary production and root biomass required investigation as it does not exist for all species in all environments (Milchunas and Lauenroth, 1993). Trampling of the vegetation may also have considerable influence on the plant community. Chandrasekara and Frid (1996) observed that trampling by humans in saltmarshes led to distinct changes in the community similar to those recorded for grazing. In addition, compaction and trampling alters the sediment density which can affect the structure and function of the benthic community (Morris and Jensen, 1998).

At Arne, the macro-infauna were most abundant in the grazed areas and least abundant in the fenced exclosures. High levels of root biomass have been found to inhibit the development of a diverse infaunal community (Hedge and Kriwoken, 2000). However, the same authors reported higher levels of invertebrates in vegetated saltmarshes than on the adjacent mudflats so clearly root biomass is only one of a number of important variables. None of the infauna species in this study are important food sources for Redshank as it feeds largely on *Corophium* spp. (Ferns, 1992). However, the infaunal species present are important food sources for other birds. For example, *Hydrobia* are consumed by Shelduck (*Tadorna tadorna*) and other species of duck (Cadee, 1994) and *Gammarus* is an important food source for waders such as Dunlin (Verkuil *et al.*, 1993), as well as for many fish that are themselves bird prey (Bartlett, 1996).

In conclusion, this study has found that deer grazing has both beneficial and detrimental effects on the saltmarsh, related to the intensity of grazing. Heavily grazed areas of the upper marsh are a concern as they lack the above ground vegetation structure to support breeding birds such as Redshank that require tussocks of tall grass. Such areas do, however, harbour a large abundance of infauna bird prey. Some grazing of the saltmarshes is clearly beneficial in terms of conservation management. Although grazing by Sika Deer is harder to manage than cattle grazing, their greater agility at crossing creeks may make them worth considering as a tool for grazing inaccessible areas of marsh.

Acknowledgements

We thank all staff at RSPB Arne, particularly Dr John Underhill-Day, Neil Gartshore and Ian Clowes for their many contributions to this work. We would also like to acknowledge Vanessa Penny, Lisa Mack and Craig House and thank them for the valuable preliminary studies they carried out by during their time as undergraduates at Bournemouth University.

References

Allport, G., O'Brien, M. and Cadbury, C. J. (1986) *Survey of Redshank and Other Breeding Birds on Saltmarshes in Britain 1985. CSD Report*, No. 649. Peterborough: Nature Conservancy Council.

Andresen, H., Bakker, J. P., Brongers, M., Heydeman, B. and Irmler, U. (1990) Long-term changes of salt marsh communities by cattle grazing. *Vegetatio*, **89**: 137–148.

Bakker, J. P. (1978) Changes in a salt-marsh vegetation as a result of grazing and mowing – a five year study of permanent plots. *Vegetatio*, **38**: 77–87.

Bakker, J. P. (1985) The impact of grazing on plant communities, plant populations and soil conditions on salt marshes. *Vegetatio*, **62**: 391–398.

Bakker, J. P. and Ruyter, J. C. (1981) Effects of five years of grazing on a salt-marsh vegetation. *Vegetatio*, **44**: 81–100.

Bartlett, C. M. (1996) Morphogenesis of *Contracaecum rudolphii* (Nematoda: Ascaridodea). a parasite of fish-eating birds, in its copepod precursor and fish intermediate hosts. *Parasite-Journal de la Société Française de Parasitologie,* **3** (4): 367–376.

Belsky, A. J. (1987) The effects of grazing: confounding of ecosystem, community and organism scales. *The American Naturalist*, **129**: 777–783.

Bertness, M. D. (1991) Interspecific interactions among high marsh perennials in a New England salt marsh. *Ecology*, **72**: 125–137.

Bouchard, V., Tessier, M., Digaire, F., Viver, J. P., Valery, L., Gloaguen, J.-C. and Lefeuvre, J. C. (2003) Sheep grazing as management tool in western European saltmarshes. *Comptes Rendus Biologies*, **326**: 148–157.

Cadee, G. C. (1994) Eider. Shelduck and other predators, the main producers of shell fragments in the Wadden Sea – paleoecological implications. *Palaeontology*, **37**: 181–202.

Cadwalladr, D. A., Owen, M., Morley, J. V. and Cook, R. S. (1972) Wigeon (*Anas penelope* L.) conservation and salting pasture management at Bridgwater Bay National Nature Reserve, Somerset. *Journal of Applied Ecology*, **9**: 417–425.

Chandrasekara, W. U. and Frid, C. L. J. (1996) Effects of human trampling on tidalflat infauna. *Aquatic Conservation: Marine and Freshwater Ecosystems*, **6**: 299–311.

Clark, K. (1993) Non-parametric multivariate analyses of changes in community structure. *Australian Journal of Ecology*, **18**: 117–143.

De Leeuw, J. and Bakker, J. P. (1986) Sheep grazing with different foraging efficiencies in a Dutch mixed grassland. *Journal of Applied Ecology*, **23**: 781–793.

Esselink, P., Zijlstra, W., Dijkema, K. S. and van Diggele, R. (2000) The effects of decreased management on plant-species distribution patterns in a salt marsh nature reserve in the Wadden Sea. *Biological Conservation*, **93**: 61–76.

Ferns, P. (1992) *Bird Life of Coasts and Estuaries*. Cambridge: Cambridge University Press.

Hedge, P. and Kriwoken, L. K. (2000) Evidence for effects of *Spartina anglica* invasion on benthic macro-fauna in Little Swanport estuary, Tasmania. *Australian Ecology*, **25**: 150–159.

Horwood, M. T. and Masters, E. H. (1981) *Sika Deer*. Second (revised) Edition. Hampshire: The British Deer Society.

Jensen, A. (1985) The effect of cattle and sheep grazing on salt-marsh vegetation at Skallingen, Denmark. *Vegetatio*, **6**: 37–48.

Mack, L. (2001) An Investigation into the Effects of the Sika Deer Populations on the Vegetation of Arne. Unpublished B.Sc. dissertation, Bournemouth University.

Mann, J. C. E. and Putman, R. J. (1989) Patterns of habitat use and activity in British populations of sika deer of contrasting environments. *Acta Theriologica*, **35**: (5): 83–96.

Michell, B., Staines, B. W. and Welch, D. (1977) *Ecology of Red Deer: A Research Review Relevant to their Management in Scotland*. Banchory: Institute of Terrestrial Ecology.

Milchunas, D. G. and Lauenroth, W. K. (1993) Quantitative effects of grazing on vegetation and soils over a global range of environments. *Ecological Monographs*, **63**: 327–366.

Mitchell-Jones, T. and Kirby, K. (1997) *Deer Management and Woodland Conservation in England.* Peterborough: English Nature.

Morris, J. T. and Jensen, A. (1998) The carbon balance of grazed and non-grazed *Spartina anglica* saltmarshes at Skallingen, Denmark. *Journal of Ecology*, **86**: 229–242.

Norris, K., Brindley, E., Cook, T., Babbs, S., Forster-Brown, C. and Yaxley, R. (1998) Is the density of redshank *Tringa totanus* nesting on saltmarshes in Great Britain declining due to changes in grazing management? *Journal of Applied Ecology*, **35**: 621–634.

Norris, K., Cook, T., O'Dowd, B. and Durdin, C. (1997) The density of redshank *Tringa totanus* breeding on the salt-marshes of the Wash in relation to habitat and its grazing management. *Journal of Applied Ecology*, **34:** 999–1013.

Pehrsson, O. (1988) Effects of grazing and inundation on pasture quality and seed production in a salt marsh. *Vegetatio*, **74**: 113–124.

Price, B. (1997) *A Survey of Saltmarsh Communities within Poole Harbour, Dorset for Breeding Waders and Gulls.* Sandy: RSPB.

Putman, R. J. (2000) *Sika Deer.* London/Fordingbridge: The Mammal Society/British Deer Society.

Ratcliffe, P. R. (1987) Distribution and current status of Sika deer, *Cervus nippon*, in Great Britain. *Mammal Review*, **17**: 39–58.

Seliskar, D. M. (2003) The response of *Ammophila breviligulata* and *Spartina patens* (Poaceae) to grazing by feral horses on a dynamic mid-Atlantic barrier island. *American Journal of Botany*, **90**: 1038–1044.

Verkuil, Y., Koolhaas, A. and Vanderwinden, J. (1993) Wind effects on prey availability – how northward migrating waders use brackish and hypersaline lagoons in the Sivash, Ukraine. *Netherlands Journal of Sea Research*, **31** (4): 359–374.

Whitehead, G. K. (1964) *The Deer of Great Britain and Ireland.* London: Routledge and Kegan Paul.

15. Sika Deer Trampling and Saltmarsh Creek Erosion: Preliminary Investigation

Craig House, Vincent May and Anita Diaz

School of Conservation Sciences, Bournemouth University, Talbot Campus, Fern Barrow, Poole, Dorset BH12 5BB

A one-year investigation into the effect of trampling by Sika Deer *Cervus nippon* on the erosion of saltmarsh creek sides compared the rates of erosion in creek sections crossed by deer with those in control areas. Significantly greater erosion of creek sides was detected in the areas where deer crossed the creeks. The greater the frequency of deer crossings (as measured by number of footprints), the greater was the rate of erosion. Erosion rates were related to loss of creekside vegetation, particularly *Puccinellia maritima*. The methods developed during the investigation provide a basis for further investigation.

Introduction

A distinctive feature of saltmarshes is the often complex system of drainage channels, termed creeks (Long and Mason, 1983) that play a vital role in marsh ecology, hydrology and nutrient dynamics (Gosselink and Mitsch, 1986; Ranwell, 1972). Any change in creek size may alter water flow velocity (Wang *et al.*, 1999) with a concomitant alteration in material transport and sedimentation characteristics of the creek (Adam, 1990).

Large populations of introduced Sika Deer *Cervus nippon* use saltmarshes around Poole Harbour. Their agility means that they can easily jump across creeks only a few metres wide and thus access large areas of marsh. A comparison of aerial photography of the saltmarsh creeks suggested that between 1971 and 1998, the areas most heavily grazed by deer had lost many of the first-order creek channels, whereas minimally grazed areas typically showed an increase in first-order channels. Although there are difficulties with the assessment of changes in area between images (Shi *et al.*, 1995), the loss of defined creek channel is consistent with the proposition that trampling is modifying the channel network.

This chapter describes an investigation into whether trampling by Sika Deer crossing creeks has an impact on the rate of erosion of the creek sides. To achieve this, the following factors were measured:

(i) the rates of erosion of creek sides in areas where deer cross the creeks and in control areas where no deer cross
(ii) the relationship between rates of erosion and number of deer footprints
(iii) the relationship between rates of erosion and abundance of creekside vegetation.

Methods

The study site was an area of saltmarsh between Shipstal Point and Gold Point, Arne. Creek sections crossed by deer were located by following deer tracks across the marsh. Fifteen 1 m sections of creek crossed by deer and 15 control sections (i.e. not crossed by deer) were chosen at random from areas across the study site. The rate of erosion of each creek was measured between February 1999 and February 2000 by comparing the cross-sectional area of each creek section using the datum technique (Lawler, 1993). From this, the percentage area newly eroded between February 1999 and February 2000 was calculated.

The usage by deer of each crossing was measured in three ways.

(i) Footprint counting. Prints were found to last normally at least 7 days so the count was achieved by smoothing away all existing prints in each area and then returning a week later to record the number of prints deposited over that week. An estimate was made of the number and orientation of deer prints within 1 m x 1 m areas immediately either side of each creek-crossing every 2–3 weeks between March 1999 and March 2000.

(ii) Trampleometer. Bayfield's (1971) method for monitoring walker pressure on footpaths was adapted for use in the marsh. An array of 66 nails was embedded into the mud at each site with a spacing of 10 cm and covering an area of 0.5 m x 1 m. Each nail had a loop of 5-amp fuse wire soldered to its head. To set the trampleometer, all loops were pulled up and it was then left in place for a week. After this time, a record was made of the number of flattened loops.

(iii) Infra-red gate. An attempt was made to calibrate both of the above indices against actual counts of number of deer crossing the creek by setting up an infra-red beam and counting the number of times it was broken. However, comparison of deer movements with the infra-red counter records showed that deer were frequently not recorded and so this method was abandoned.

The percentage cover of each plant species within the 1 m x 1 m areas immediately either side of each creek-crossing was recorded in February 1999 and February 2000.

Results

Creek sections crossed by deer were found to be significantly more eroded than control creeks ($t = -3.40$, $P = 0.002$) (Figure 1). A significant correlation was found between erosion and the number of footprints pointing across the creek (Spearman rank $r = 0.518$, $P = 0.048$). No significant relationship was detected using the trampleometer (Spearman rank $r = 0.469$, $P = 0.53$), possibly because deer sometimes detected and avoided the trampleometer.

Creek sections crossed by deer were significantly less vegetated than control creeks ($t = -7.49$, $P < 0.001$) (Figure 2). Further analysis showed that, at the level of individual plant species in creeks crossed by deer, the loss of cover of *Puccinellia maritima* was significantly correlated with increased erosion (Spearman rank $r = -0.527$, $P = 0.043$).

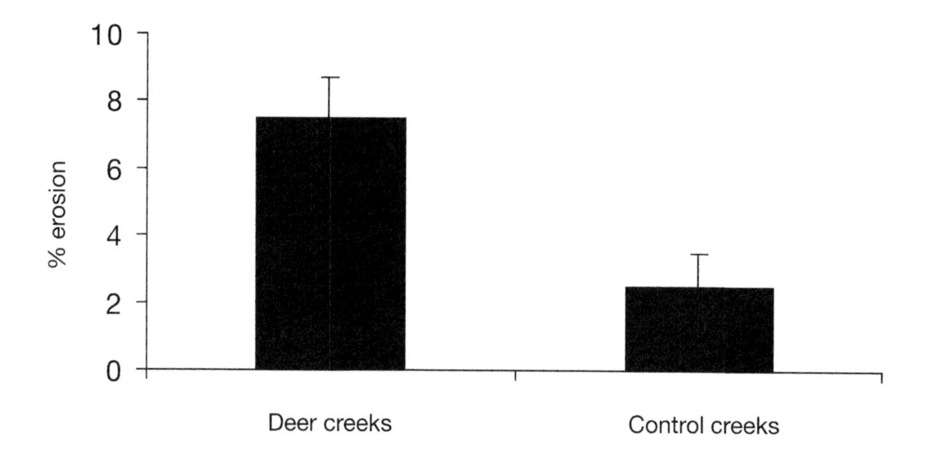

Figure 1 Mean percentage erosion in creek sections crossed by deer and in control section. Error bars show the standard error for each mean.

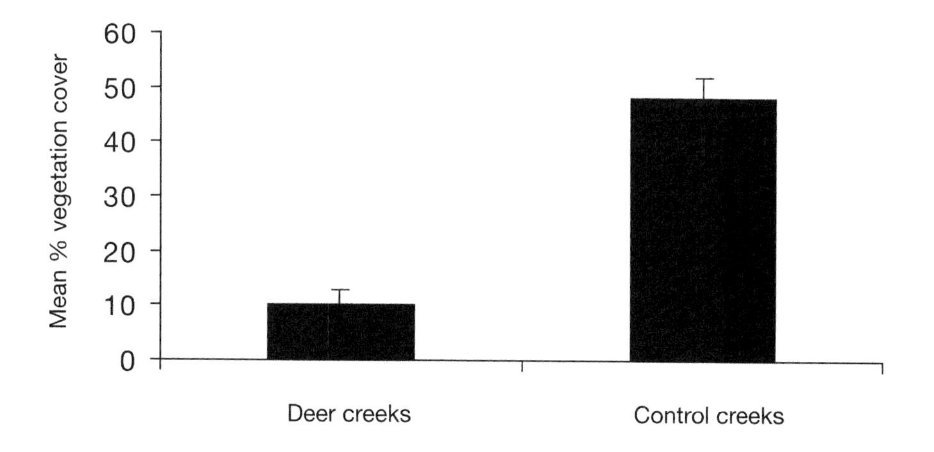

Figure 2 Mean percentage total vegetation cover in creek sections crossed by deer and in control section. Error bars show the standard error for each mean.

Discussion

The results from this preliminary study indicate that creek sections crossed by deer erode more quickly than control areas and that this is related to deer trampling. Trampling, like other forms of mechanical loading, can have direct effects on erosion rates by causing reduced soil porosity and increased compaction (Ford and Grace, 1998; Adam, 1990). Reduced porosity results in greater friction between particles decreasing sediment cohesion (Pestrong, 1965) and in turn increasing erosion rates (Shi *et al.*, 1995). Future investigations will need to consider these effects.

Trampling can also have an indirect effect on erosion via its impact on vegetation cover. In the current study, a strong relationship between increased erosion and the loss of cover of *Puccinellia maritima* was found. Plant species are well known to differ in their abilities to bind sediments and baffle tidal currents (Gabet, 1998; Garofalo, 1980; Chapman, 1974). *Puccinellia maritima* is a shallow-rooted, stoloniferous grass that is particularly effective at colonizing and stabilizing bare mud (Hubbard, 1967). Its loss may assist the erosion around creeks.

Many factors can influence creek erosion rates: this study has shown the potential for deer activity to be a contributory factor in significant localized erosion of creek sides. However, the impact of deer trampling may simply be accelerating wider trends in saltmarsh morphodynamics. Both the local scale effects of trampling and their relation to wider trends in Poole Harbour need to be explored further.

Acknowledgements

We thank Neil Gartshore and all the staff at RSPB Arne for their assistance during this investigation.

References

Adam, P. (1990) *Saltmarsh Ecology*. Cambridge: Cambridge University Press.

Bayfield, N. G. (1971) Thin wire trampleometers – a simple method for detecting variations in walker pressure across paths. *Journal of Applied Ecology,* **8**: 533–536.

Chapman, V. J. (1974) *Saltmarshes and Salt Deserts of the World.* London: Leonard Hill.

Ford, M. A. and Grace, J. B. (1998) Effects of vertebrate herbivores on soil processes, plant biomass, litter accumulation and soil elevation changes in a coastal marsh. *Journal of Ecology,* **86**: 974–982.

Gabet, E. J. (1998) Lateral migration and bank erosion in a saltmarsh tidal channel in San Francisco Bay, California. *Estuaries*, **21** (4b): 258–270.

Garofalo, D. (1980) The influence of wetland vegetation on tidal stream migration and morphology. *Estuaries*, **3** (4): 258–270.

Gosselink, J. G. and Mitsch, W. J. (1986) *Wetlands*. New York: Van Nostrand Reinhold.

Hubbard, C. E. (1967) *Grasses. A Guide to Their Structure, Identification, Uses and Distribution in the British Isles.* London: Penguin Books.

Lawler, D. M. (1993) The measurement of river bank erosion and lateral channel change: A review. *Earth Surface Processes and Landforms,* **18**: 777–821.

Long, S. P. and Mason, C. F. (1983) *Saltmarsh Ecology*. London: Blackie.

Pestrong, R. (1965) The development of drainage patterns on tidal marshes. *Geological Science,* 10 (2). Stanford: Stanford University Publications.

Ranwell, D. S. (1972) *Ecology of Saltmarshes and Sand Dunes*. London: Chapman and Hall.

Shi, Z., Lamb, H. F. and Collin, R. L. (1995) Geomorphic change of salt-marsh tidal creek networks in the Dyfi Estuary, Wales. *Marine Geology,* **128** (1–2): 73–83.

Wang, Y. P., Zhang, R. S, and Gao, S. (1999) Velocity variations in salt marsh creeks, Jiangsu, China. *Journal of Coastal Research,* **15** (2): 471–477.

The Ecology of Poole Harbour
John Humphreys and Vincent May (editors)

16. Marine Fisheries of Poole Harbour

Antony Jensen[1], Ian Carrier[2] and Neil Richardson[2]

[1]Southampton Oceanography Centre, University of Southampton, Hampshire SO14 3ZH
[2]Southern Sea Fisheries District, 64 Ashley Road, Poole, Dorset BH14 9BN

Poole Harbour is the location of a range of fishing and aquaculture activities which are worth approximately £2 million per year to the local economy. The majority of fisheries activity is controlled by the Southern Sea Fisheries District; most of the harbour lies within the boundaries of a Regulated Fishery Order. The Environment Agency is responsible for the eels fishery and any salmonid issues. The aquaculture beds produce oysters, mussels, cockles and clams from the designated Several Order, whilst mullet, cockles and clams comprise the majority of the fishery catches. The harbour is also important to recreational fishers, Flounder being the most important catch.

Introduction

Poole Harbour supports a diversity of fisheries which are managed with legislation implemented through the Environment Agency (eels and salmonids) and Southern Sea Fisheries District (SSFD) (all other marine fisheries). Fishing effort in the harbour varies with (amongst other things) fishing season, weather and the first sale price of catch. This chapter concentrates on the fisheries managed by SSFD.

The Southern Sea Fisheries District is one of 12 Sea Fisheries Districts which regulate fisheries along the coasts of England and Wales out to 6 nm. The formation of Local Sea Fisheries Committees was originally authorized by the Sea Fisheries Regulation Act of 1888 and the Southern District was instituted by a Board of Trade Order on 7 June 1893. Although the original Act was replaced by a new Sea Fisheries Regulation Act in 1966, the responsibility for managing and policing the coastal fisheries remained unchanged. In 1993, the Southern District was considerably enlarged when the limits were extended from 3 miles to 6 miles seaward of baselines and in 1995, under the Environment Act, Committees were given additional powers to safeguard the marine environment.

As the statutory authority responsible for managing and policing the coastal fisheries, the Committee is empowered to make by-laws controlling fishing including size limits, gear restrictions and seasonal limits. By-laws may also be made to protect the marine environment. The Committee's officers are also empowered to enforce a wide range of both national and EU fisheries legislation within the Southern District and these powers apply on land within the boundaries of those local authorities which contribute to the Committee's finances.

The Committee administers the Poole Fishery Order which is a hybrid Order combining a Regulated and Several Fishery (Figure 1), and has been in force since 1915. The Committee receives income from licence and lease fees which are recharged to cover the policing of these fisheries. The Several Fishery facilitates shellfish aquaculture within Poole Harbour with the SSFD leasing harbour seabed from the Crown Estate Commissioners and sub-leasing to shellfish farmers to grow oysters, clams, cockles and mussels. The Regulated Fishery covers most of the harbour and provides the regulatory framework within which commercial fisheries for oysters, mussels and clams are managed.

Fishing activity is monitored by District Fishery Officers in the harbour throughout the year. Random checks are carried out at sea and on landing locations, to ascertain whether or not the catches fished fall within the guidelines of the relevant legislation. The types of vessels inspected include both hobby and licensed fishermen. Prosecutions conducted against people removing clams from the fishery out of season have resulted in guilty verdicts.

There are approximately 98 under 10 m licensed fishing boats moored within Poole Harbour. Some of these boats fish out of the harbour as far as Mid English Channel shellfish grounds. Thirty-one of those boats hold a clam licence which allows them to fish for clams, mainly the Manila Clam *Tapes philippinarum,* within the Regulated Fishery.

Some of the methods used to fish have potentially negative impacts on important plant and animal life within the harbour. The harbour carries many conservation designations that include a Special Protection Area (SPA), a candidate Special Areas of Conservation (cSAC), a RAMSAR Site and also SSSI sites. Of particular value are eel-grass (*Zostra marina)* beds (which have been mapped), the presence of Seahorses (*Hippocampus ramulosus),* reported as having been caught in mullet nets and returned unharmed and the general intertidal invertebrate communities that provide food for wintering birds.

Fishing and aquaculture within the harbour

Several Fishery – aquaculture

The leased aquaculture beds (Figure 2) have been worked for over 50 years. The area currently under lease is approximately 182 ha and oysters, cockles, mussels and clams (Manila and Palourdes) are the species farmed. The amount of shellfish laid down on these plots is recorded as well as the amounts of shellfish harvested.

Approximately 100 tonnes of seed Edible Cockle *Cerastoderma edule* have been laid on the Several grounds each year. In the region of 2 million individual Manila Clams *Tapes philippinarum* and 2 million individual Pacific Oysters *Crassostrea gigas* are also laid each year within the Several grounds. Approximately 800–1000 tonnes of seed mussel *Mytilus edulis* are grown within the Several Fisheries at any one time. The value of landings from these beds is in excess of £1 million per year.

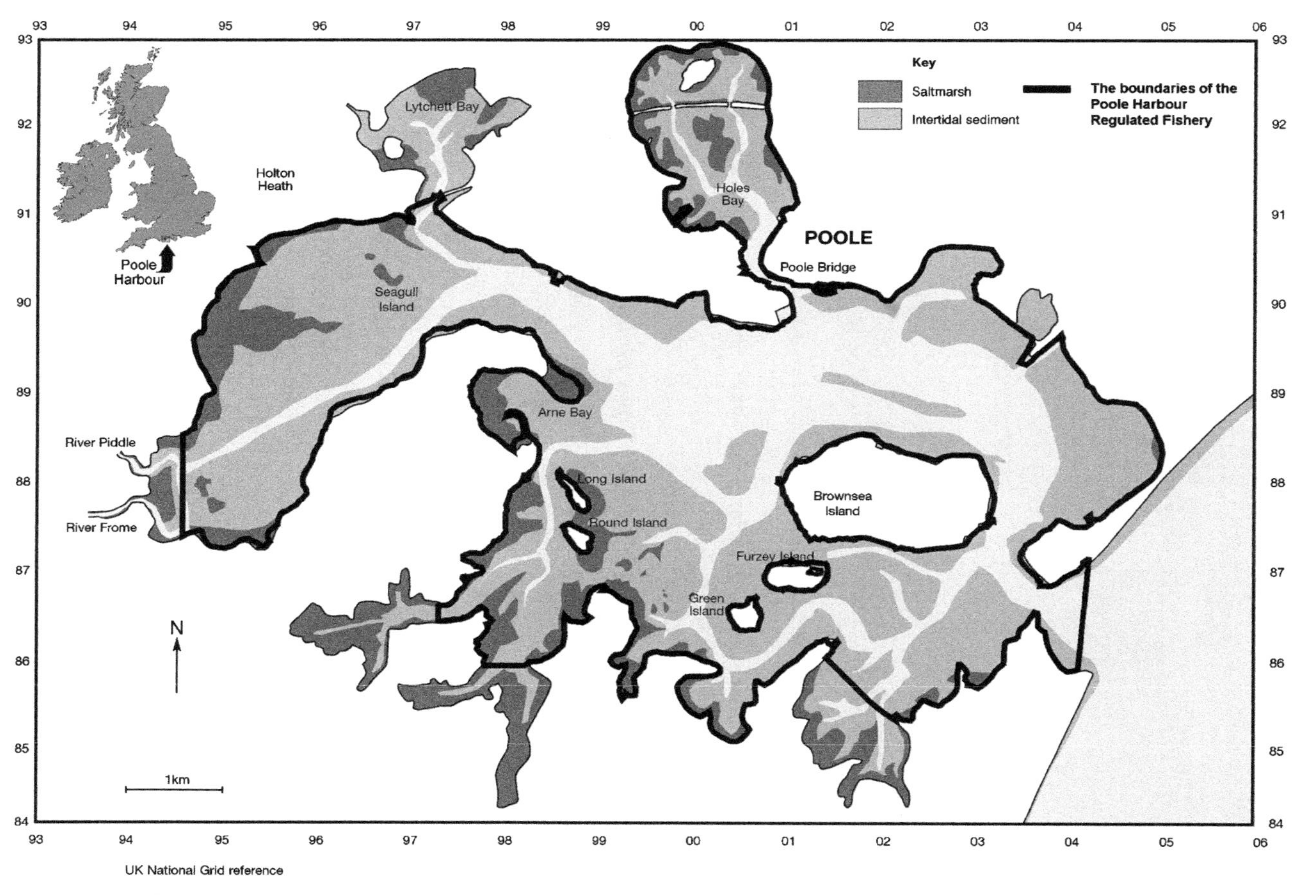

Figure 1 Area of the harbour within the Poole Fishery Order.

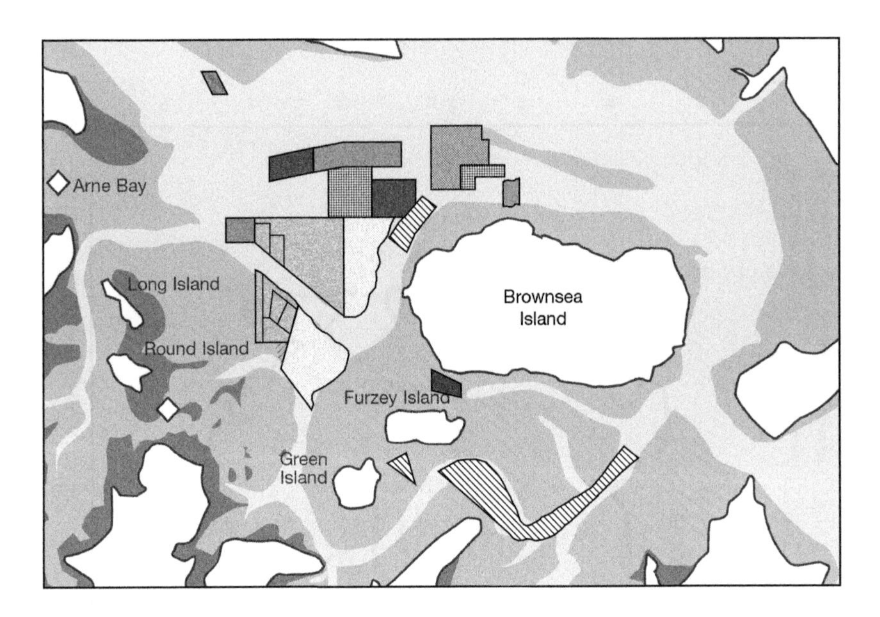

Figure 2 Area of the harbour reserved for aquaculture. Shading and hatching denotes different leaseholders.

Regulated Fishery for the Manila Clam *Tapes philippinarum*

The Manila Clam *Tapes philippinarum* is a Pacific species which was introduced to the harbour in 1989 by Othniel Shellfisheries. It was not expected to reproduce successfully but by 1992, it was obvious that the cultivated clams (sourced from a hatchery) had spawned and the spat has successfully settled and grown to a marketable size outside the aquaculture beds. This fishery is now operated within the Poole Order and 31 licences have been issued. From late October to early January, clams are removed from the fishery using a hand-held dredge or scoop which is towed along the seabed by a small under 10 m fishing vessel. The maximum size of the scoop is 460 mm wide x 300 mm high x 460 mm deep, with teeth on the front edge to a depth of 90 mm. By-laws are in place to govern the overall size of the dredge and its mesh. Gear regulations are designed to minimize disturbance and allow young spat and infauna to go through the mesh of the dredge. Fishers are obliged to sort their catch immediately it is landed on the boat and so all under minimum landing size clams and by-catch are returned to the water within minutes of arrival at the surface.

Modifications have been introduced over the years by way of water jets attached to the front entrance of the scoop to allow the mud to be moved through the basket with less physical effort (pump-scoop) (Figure 3). Ongoing investigations into the impacts of this type of fishing gear on the infaunal communities and the sediments in which they live are documented elsewhere (Jensen *et al.*, chapter 13; Cesar, 2003).

Figure 3 A pump-scoop used in the Poole clam fishery.

Approximately 300 tonnes of Manila Clams have been removed from the Regulated Fishery in the years 2001 and 2002. Areas of the harbour outside the Regulated Fishery which support clams have yielded approximately 50–100 tonnes over the same time. The clam fishery is a valuable resource to the fishers of Poole as first sale price can reach £5 per kg. Enforcement of fishing regulations is currently a major problem as the high prices that clams fetch are attracting poachers from both within and outside the fishing community. The value of landings is in excess of £1 million annually.

To support the fishery management effort in creating a sustainable, environmentally responsible fishery regular surveys are carried out to: (i) establish the density and size frequency distribution of the clams in three fishing areas; and (ii) provide data on the size and variety of the infaunal communities in both fished and unfished (or at worst lightly fished) areas of the harbour. In 2003, a Ph.D. study was completed (Grisley, 2003), which described the reproductive ecology of the naturalized Manila Clam population. Data from this have already influenced fishery decision-making in the harbour.

Cockle *Cerastoderma edule* fishery

Cockles are harvested between 1 May and 31 January each year. The method used to remove the cockles from the fishery has changed recently so that now most commercial fishermen use a trailed pump-dredge. The type of dredge has altered little over time but the pumped water spray at the mouth of the dredge (to flush sediment out of the dredge) was introduced in 2002. This technique is thought to be unique to Poole Harbour and should not be confused with hydraulic or suction dredges used in fisheries elsewhere. Cockles are subject to various restrictions through by-laws including a minimum landing size of 23.8 mm, a closed season from 1 February to 30 April, and restrictions on the type of gear and method of fishing. However, cockles are not included in the licensing of fishing section of the Poole Fishery Order, therefore, the SSFD cannot issue licences for the cockle fishery.

As cockle fishing effort increased with the use of pump technology and the ensuing noise problem from the engine-driven pumps (now minimized following the issue of noise abatement notices), environmental concerns were raised by English Nature and RSPB about the impact of this fishing technique on the sediment granulometry and invertebrate communities of the cockle beds. Infauna and sediments were monitored over the summer of 2003 (Parker, 2003) and these results reported by Parker and Pinn (see chapter 17).

In contrast to the efficiency of the boat-based cockle fishery, there is an artisanal, hand-raking fishery that harvests cockles (and clams in some parts) on the sandier littoral areas of the harbour, especially Whitley Lake. Collecting cockles in this fashion for a 'feed for the family' has been a harbour tradition for generations. The cockle population has been sufficiently robust in some areas such as Whitley Lake to support both artisanal hand pickers and efficient pump-dredge fishers.

American Hardshell Clams (*Mercenaria mercenaria*)

This species of clam was fished in Southampton Water in the 1970s until the mid-1980s after being introduced from America in the 1920s–30s. The main market was abroad and Hardshell Clams were laid up on the seabed in Poole (and other places) while waiting for shipment. While they were laid on the seabed, the clams reproduced and their spat settled and survived. This species has been found in the harbour ever since. They are not found in great numbers but hold a great commercial value and are subject to a minimum landing size of 6.3 cm. They can also be found in the Solent Regulated Fishery where they have a closed season from 1 November to the last day in February in the following year both days inclusive.

European Oysters (*Ostrea edulis*)

Oysters are naturally occurring within the harbour. They were laid down as seed on some of the Several shellfish beds some years ago and oysters from the Solent fishery were relaid to grow on for a year before harvest, but stocks were reduced significantly by an outbreak of bonamia in the 1980s. Oysters have been farmed since then but never in such densities and bonamia is still present in the harbour.

Grey Mullet (*Chelon labrosus*)

The netting for Grey Mullet is unique to this area with the use of a traditional Poole canoe and a ring net laid in a decreasing circular pattern. The net is hand-hauled over the transom of the canoe and the fish taken out as the net is recovered. This type of netting takes place from March through to November with approximately five commercial fishermen and five hobby fishermen carrying out this practice. Grey Mullet is subject to a (by-law) minimum landing size of 30 cm.

Flounder (*Platichthys flesus*)

Trawling in the harbour has taken place for many generations with small trawls fitted on under 10 m boats to catch the Flounder. This species is the focus of much recreational angling effort within the harbour and is said to be one of the biggest Flounder fisheries in the UK. This fishery is a magnet for anglers from October to January during the Flounder season. Flounders are subject to a (by-law) minimum landing size of 27 cm.

Bass (*Dicentrarchus labrax*)

Bass is targeted in the same way as Grey Mullet except the Bass fishery has a closed season covering a majority of the harbour. This is a Bass Nursery Area (National Legislation: SI 1990 No. 1156) and is closed to the removal of any Bass by any vessel from 1 May to the 31 October of each year. The prohibition on Bass fishing in the nursery area does not apply to fishing from shore. Bass are subject to a minimum landing size of 36 cm and this applies to both commercial fishermen and anglers.

Lesser Sand-eel (*Ammodytes tobianus*)

Sand-eel netting takes place on a few of the sub-littoral sand banks to the south of the Middle Ship Channel. This type of fishing activity is carried out during the spring and summer months and is important to the fishery as it supports the commercial and recreational rod anglers by providing bait for the fishing of Bass outside the nursery area.

The Sand-eel is fished using a trawl net with a small mesh size. A by-catch is allowed dependent on the mesh size of the net.

Prawns

There is a prawn fishery but it does not have a great commercial importance, although some expansion of effort is thought to be possible given the stocks. The prawn fishery is closed from 1 January to 31 July of each year by a by-law that is specific to Poole Harbour.

Eels

Taken mainly in fyke nets and controlled by the Environment Agency, eel fishing in the harbour is of local importance. The current main regulatory issue is the importance of fitting Otter guards to the nets to prevent Otters from entering to eat the eels and then drowning as they cannot get out of the complex nets.

Charter fishing boats

The charter fleet is one of the biggest in the UK with 35 boats in total. The fleet works within the constraints of the harbour during bad weather. This type of activity attracts recreational anglers to the area from all over Britain.

Angling clubs and associations

There are approximately eight sea angling clubs in the Poole area alone. Members of these and other local clubs from neighbouring towns and villages use the shoreline to carry out sport fishing. The Flounder fishery is one of the largest attractions for anglers around Britain (see above).

Bait dragging

Poole Harbour supports a number of boats that harvest ragworm from areas of the harbour using a unique two pronged 'dredge', which is dragged over the seabed removing the worms from the upper layers of the seabed. The ragworm is sold as bait for recreational anglers. Agreements between the 'bait draggers' and other fishing interests in the harbour minimize 'use of seabed' conflicts. Bait dragging is unique to Poole Harbour and lies outside the fishing regulations that are in force.

Summary

Poole Harbour is an important aquaculture and fishery resource. It provides sheltered moorings for the fishing fleet and a productive environment for fishers in poor weather conditions, giving access to (depending on season) finfish and bivalve fisheries. The seabed of the harbour supports a significant bivalve aquaculture industry for oysters, mussels, Manila Clams, Palourdes and cockles.

This chapter is intended to be informative but should not be used as a definitive statement of current regulations.

References

Cesar, C. P. (2003). The Impact of Clam Fishing Techniques on the Infauna of Poole Harbour. Dissertation submitted in partial fulfilment of the requirements for the degree of M.Sc. (Oceanography), School of Ocean and Earth Science, University of Southampton.

Grisley, C. (2003) The Ecology and Fishery of *Tapes philippinarum* (Adams and Reeve, 1850) in Poole Harbour, UK. A thesis presented for the degree of Doctor of Philosophy, School of Ocean and Earth Science, University of Southampton.

Parker, L. C. (2003) An investigation into the ecological effects of pump scoop dredging of cockles (*Cerastoderma edule* L.) on the intertidal benthic communities in Poole Harbour. A research project and working paper submitted as part of the requirements for MSc/PGDip coastal zone management, School of Conservation Sciences, Bournemouth University.

John Humphreys and Vincent May (editors)

17. Ecological Effects of Pump-scoop Dredging for Cockles on the Intertidal Benthic Community

Linda Parker and Eunice Pinn*

School of Conservation Sciences, Bournemouth University, Talbot Campus, Fern Barrow, Poole, Dorset BH12 5BB

*Now at the Joint Nature Conservation Committee, Dunnet House, 7 Thistle Place, Aberdeen AB10 1UZ. Fax: 01224 621488, e-mail address: eunice.pinn@jncc.gov.uk

Awareness of the ecosystem effects of fishing activities on the marine environment means that there is a vital need to assess the direct and indirect effects of those activities that may have negative effects on target and non-target species. The Edible Cockle *Cerastoderma edule* is the target of an artisanal and commercial fishery that occurs in estuarine and intertidal habitats across northern Europe. Poole Harbour has opened up its cockle beds to pump-scoop dredging over the last few years. This study investigated the effect of pump-scoop dredging on the intertidal sedimentary environment and the macro-infaunal community. The results demonstrated that the dredging did not have an effect on the size distribution of sediment particles. After the fishery opened, no immediate impact of cockle dredging on the infaunal community was observed. Within 3 months, however, a reduction in species richness and abundance of the benthic community was noted. This may be indicative of a chronic rather than acute impact. However, further investigations are required to assess whether this was directly related to pump-scoop dredging. The findings of this study are discussed in relation to possible impacts on the ecosystem as a whole.

Introduction

One of the most pressing issues in coastal zone management is how to accommodate the wide range of uses and activities in the coastal margin such that the ecology of intertidal and nearshore marine habitats is protected (Kaiser *et al.*, 2001; Hiddink, 2003). In particular, fisheries management needs to consider both environmental and political sensitivities in coastal habitats owing to the extractive nature of harvesting processes, disturbance to the habitat and potential conflicts between multiple users (Kaiser *et al.*, 2001; Atkinson *et al.*, 2003).

Harvesting of the Edible Cockle (*Cerastoderma edule*) has taken place in Britain at least since mediaeval times (Rostron, 1995). Traditionally, cockles have been harvested from intertidal sediments at low tide, using hand rakes and riddles, the latter used to separate out undersized individuals (Jenkins, 1991). More recently, harvesting has become mechanized with the use of hydraulic dredges (Hall *et al.*, 1990). Due to its greater efficiency, a hydraulic dredge can remove a larger number of cockles from an area than several men raking for the same length of time (Pickett, 1973). Thus, a ground can be commercially 'fished-out' by dredges in a much shorter time and the cockle densities remaining are usually much less than those left after raking (Pickett, 1973).

Nearly all commercial concentrations of cockles occur on intertidal flats in lower reaches of estuaries. They live buried beneath the surface of fine and very fine sands and must be dislodged from sediments to be collected (Coffen-Smout, 1998). Consequently, all methods of gathering cockles inevitably involve mechanical disturbance of the habitat (Rees, 1996). This can have a variety of effects depending on the ambient community, area disturbed and fishing pressure, all of which vary among different fisheries and digging areas (Brown and Wilson, 1997). The main concerns regarding hydraulic dredging are that it might cause excessive stock depletion, impair future spatfall, and be damaging ecologically for both the benthos and birdlife (Broad, 1997; Cotter *et al.*, 1997; Hall and Harding, 1998; Bradshaw *et al.*, 2002). Consequently, harvesting may have an adverse effect on ecosystem functioning, leading to long-term changes in community structure (Gaspar *et al.*, 2002; Hiddink, 2003).

Throughout north-west Europe, estuarine habitats are home to internationally important populations of shorebirds and are also of local economic importance, due to substantial stocks of commercially fished shellfish (Atkinson *et al.*, 2003). Poole Harbour is no exception to this general rule. In its shallow waters, fishing activity is intense, including pump-scoop dredging for cockles (N. Richardson, pers. comm.). Pump-scoop dredging involves pumping seawater through a dredge to release the shellfish from the sediment before it is scooped up. In Poole Harbour, the cockle fishery is unlicensed and is, therefore, difficult to control (N. Richardson, pers. comm.). This has led to widespread concern about the impact that dredging may have on both the cockle stocks and other non-target species inhabiting the same habitat. The aim of this study was to assess the impact of pump-scoop dredging on the non-target intertidal macro-infauna.

Materials and methods

Two sites located within the Whitley Lake area of Poole Harbour were chosen for this study. Site A was to the east of Salterns Quay and Site B was situated east of the pier of the East Dorset Sailing Club based at Evening Hill, on Sandbanks Road. The substratum in the study area is mostly sandy mud with some patches of shingly ground close inshore. Both sites have a relatively flat and uniform topography. A constraint of the study was the lack of a suitable control site with the same environmental conditions but no cockle dredging. However, sampling did start before the cockle fishery opened and the work was conducted over the period of the year where the macro-faunal abundance and diversity would be expected to be highest (Souza and Gianuca, 1995; Tuya *et al.*, 2001; Rueda and Salas, 2003).

At each site, samples were collected in April before the cockle fishery season opened, and then again in May, June and July during the season. Every month, samples were collected from five positions at each site. Each position was at least 25 m from the next, in a line approaching low water. To ensure that no area was resampled and that trampling effects of sampling were kept to a minimum, each month the sampling position was moved transversely along the shore, at least 2 m from the previous position.

To examine changes in sediment particle size composition that might occur due to the pump-scoop dredging, one core sample (measuring 100 mm x 100 mm x 100 mm) was collected each month from each sampling position. Before sieving, the samples were oven dried at 65 °C. Once sieved, the degree of sorting, skewness and kurtosis (a measure of the peakedness of the distribution) were calculated. Changes in size distribution were analysed using Kruskal-Wallis tests (SPSS).

At each position, four sediment cores (measuring 200 mm x 200 mm x 100 mm depth) were collected at random for the infaunal survey. The cores were washed *in situ* over a 1 mm sieve. The residue was taken to the laboratory, where each sample was again washed over a 1 mm sieve and the remaining residue was preserved in 70% alcohol. The infauna was counted and identified to species level where possible. Species diversity was compared and contrasted using a variety of indices, including Margalef's Index (d), the Shannon Index (H') and Pielou's Evenness Index (J). These diversity indices were chosen as they are the most commonly used. Differences were tested statistically using analysis of variance (ANOVA) (SPSS). Prior to analysis, no significant differences were found between the individual quadrats for any single month and, therefore, these data were combined. In addition, the data were tested for homogeneity of variances using Levene's test. After ANOVA, Tukey HSD *post hoc* tests were conducted to assess where there were variations between months.

The infaunal data were also analysed using standard multivariate techniques with the package PRIMER (Plymouth Routines In Multivariate Ecological Research; Clarke, 1993). The Bray-Curtis similarity measure was used to calculate similarities among observations for cluster analysis and multidimensional scaling (MDS). Similarity percentages (SIMPER) were used to determine the characterizing taxa (i.e. accounted for 75% of the similarity) (Clarke, 1993). Variations in the abundance of these characterizing fauna were analysed using ANOVA (SPSS). No significant differences were found between the individual quadrats for any single month and, therefore, these data were combined for the remaining analyses. Prior to ANOVA, data were tested for homogeneity of variances using Levene's test. Heterogeneous data were square root transformed (Underwood, 1997). This, however, did not remove the heterogeneity. Underwood (1997) reported that for large balanced data sets, problems associated with violations of assumption of homogeneity and normality are unlikely to affect the F ratio. It was, therefore, decided to undertake the ANOVA using the non-transformed data, but with a more conservative probability of 0.01 (Underwood, 1997).

Results

Sediment results

The sediment from both sites was predominantly fine sand (Table 1). There was little change in the size distribution of the sediment on a monthly basis. No significant differences were observed for either site (Site A: K = 1.26, $P > 0.05$; Site B: K = 1.02, $P > 0.05$). In general, the sediment was found to be fine sorted and very leptokurtic (i.e. more peaks in the centre and tails of the size distribution relative to the normal distribution) (Table 1).

Community results

Prior to dredging, both sites exhibited a similar species richness (Site A: 17.2 ± 1.1; Site B: 17.0 ± 2.3) and total number of individuals (Site A: 412.8 ± 133.3; Site B: 439.6 ± 226.9). Three months after dredging began, species richness had declined (Site A: 12.6 ± 0.9; Site B: 14.8 ± 2.3) (Figure 1a). This reduction was found to be statistically significant (Table 2), with *post hoc* tests revealing significant differences between the July data and all other months. The reduction in the total number of individuals with the onset of dredging was more pronounced (Site A: 238.2 ± 84.6 per core; Site B: 216.8 ± 92.6 per core), declining by 42.3% at Site A and 50.6% at Site B (Figure 1b). This decline was found to be statistically significant (Table 2), with Tukey's test indicating April and July were significantly different.

Margalef's Index varied between the two sites. At site A, there was a general decline in the index whilst at Site B, there was an initial increase, followed by a decline (Figure 1c). These differences were found to be significant for both month and site (Table 2). *Post*

Table 1 Sediment analysis

	Site A (weight [g])				Site B (weight [g])			
	April	May	June	July	April	May	June	July
Modal grain size (mm)	0.212	0.212	0.212	0.212	0.212	0.212	0.212	0.212
Median grain size (mm)	0.168	0.512	0.552	0.516	0.151	0.278	0.156	0.123
Mean grain size (mm)	0.452	0.510	0.527	0.508	0.473	0.496	0.459	0.458
Sorting	0.462	0.453	0.451	0.465	0.486	0.448	0.464	0.478
Kurtosis	-2.254	-2.218	-2.259	-2.277	-2.393	-2.289	-2.252	-2.347
Skewness	0.253	0.185	0.154	0.186	0.199	0.167	0.269	0.234

a) Species richness

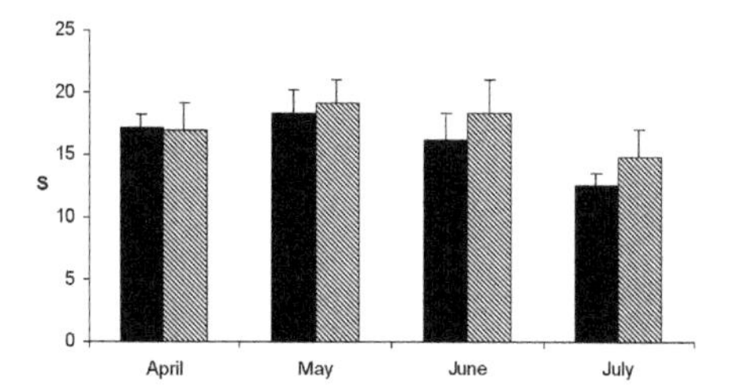

b) Individual abundance

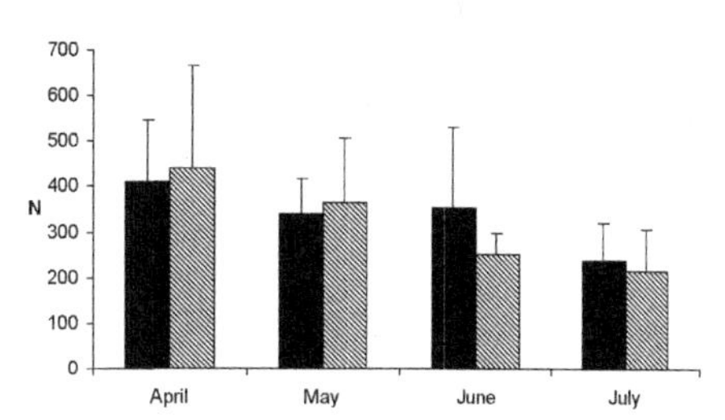

c) Margalef's Index

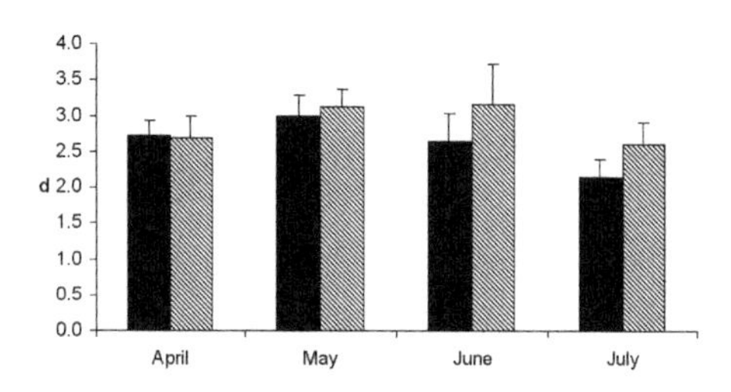

d) Pielou's Evenness Index

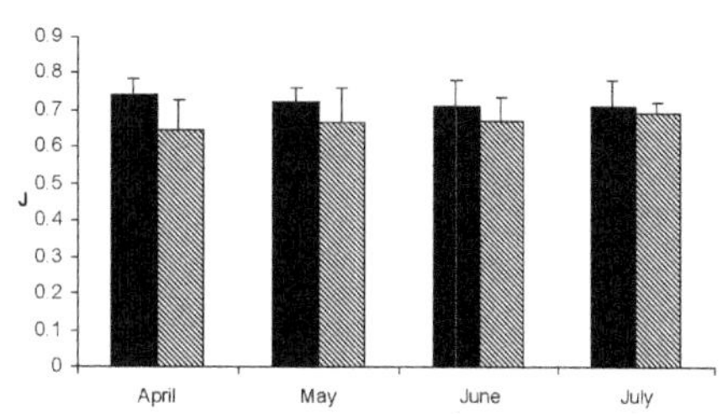

e) Shannon Index

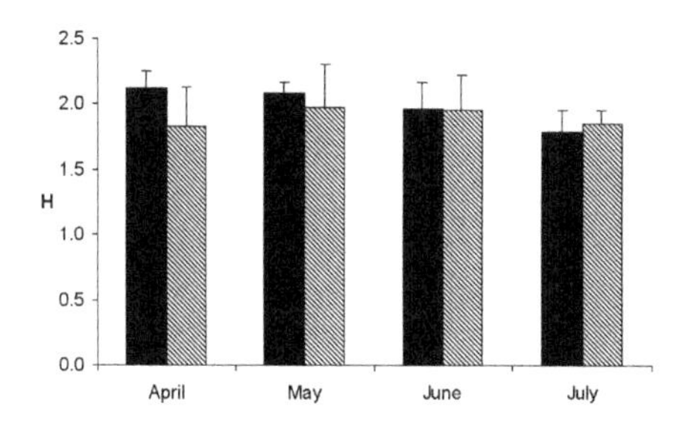

Figure 1 Changes in diversity as a response to cockle dredging (Site A: solid bars; Site B: striped bars).

hoc tests showed May and June to be significantly different from July. Little change was observed in Pielou's Evenness Index for Site A, whilst a slight increase was observed for Site B (Figure 1d). Statistical analysis showed significant differences between the two sites, but not between the different months (Table 2). Shannon's Index showed a general decline for both sites (Figure 1e), which was not found to be significant.

Table 2 Diversity ANOVA results

	Species richness		Total number of individuals			
	F	*P*	F	*P*		
Site	4.006	ns	0.195	ns		
Month	11.904	<0.001	3.863	<0.05		
Site * Month	0.878	ns	0.495	ns		
	Margalef's Index		**Pielou's Evenness Index**		**Shannon Index**	
	F	*P*	F	*P*	F	*P*
Site	6.236	<0.05	6.236	<0.05	1.682	ns
Month	8.057	<0.001	0.032	ns	1.687	ns
Site * Month	1.606	ns	0.727	ns	1.230	ns

ns – not significant at $P = 0.05$.

Cluster analysis of the abundance data identified the presence of four clusters or groups at a similarity level of 50% (Figure 2), which were also distinguished by multidimensional scaling. Group 1 consisted of all samples collected at Site B during April, May and June. Group 2 consisted of samples collected from Site A during April, May and June. Group 3 consisted of samples collected in July at Site A and Group 4 contained the Site B July samples.

Effects on individual species

SIMPER was used to identify the characterizing species at each site and for each month. The characterizing species at Site A, accounting for 80% of the similarity, were *Cingula trifasciata* (a gastropod mollusc), *Scoloplos armiger* (a polychaete worm), *Hydrobia* spp. (spire shells) and *Arenicola marina* (lugworm). At Site B, they were *S. armiger*, *A. marina*, *C. trifasciata*, *Corophium* spp. (an amphipod crustacean) and *Urothoe* spp. (an amphipod crustacean). The characterizing species for April, accounting for 80% of the similarity, were *S. armiger*, *Cingula trifasciata* and *Hydrobia* spp. May and June were similar to April, with the addition of *A. marina*. In July, *Urothoe* spp., *C. trifasciata*, *A. marina* and *Corophium* spp. were identified as characterizing species. Additional individual investigations were, therefore, conducted for: *S. armiger*, *Cingula trifasciata*, *A. marina*, *Hydrobia* spp., *Corophium* spp. and *Urothoe* spp.

The most obvious change in abundance throughout the duration of the sample collection was observed in *S. armiger*. Prior to dredging in April, both sites contained this species. However, with the onset and continuation of dredging, abundance decreased to zero in July at both sites (Figure 3a). Using a two-way ANOVA, these differences were found to be significant (Table 3), with Tukey's HSD test identifying April as being significantly different from June and July.

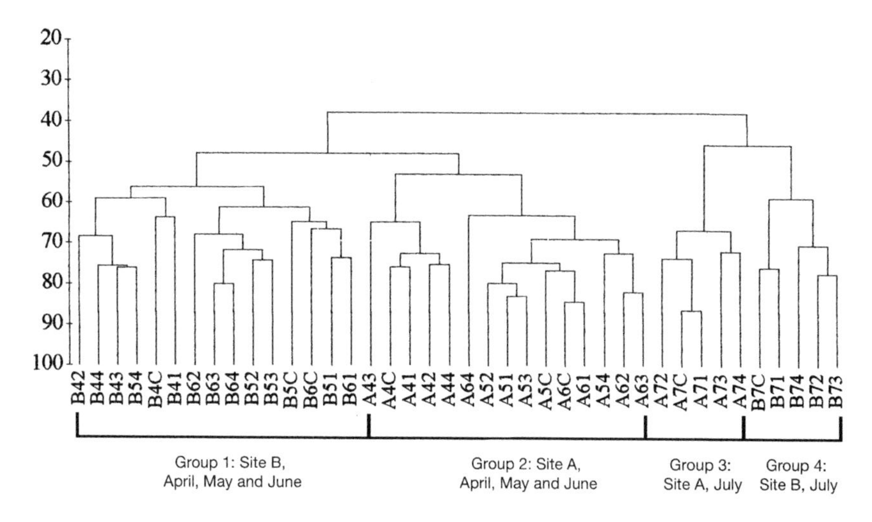

Figure 2 Cluster analysis results. Core coding: the letter refers to the site, the first number represents the month (4 – April, 5 – May, 6 – June, 7 – July) and the second, the sampling position.

Table 3 Species ANOVA results

	Scoloplos armiger		***Cingula trifasciata***		***Arenicola marina***	
	F	***P***	**F**	***P***	**F**	***P***
Site	15.184	<0.001	13.868	<0.001	6.755	<0.01
Month	57.919	<0.001	4.814	<0.01	4.144	<0.01
Site * Month	3.020	ns	3.674	ns	1.852	ns
	***Hydrobia* spp.**		***Corophium* spp.**		***Urothoe* spp.**	
	F	***P***	**F**	***P***	**F**	***P***
Site	77.201	<0.001	51.616	<0.001	58.349	<0.001
Month	5.553	<0.001	1.545	ns	33.540	<0.001
Site * Month	3.391	ns	1.520	ns	2.000	ns

ns – not significant at $P = 0.05$.

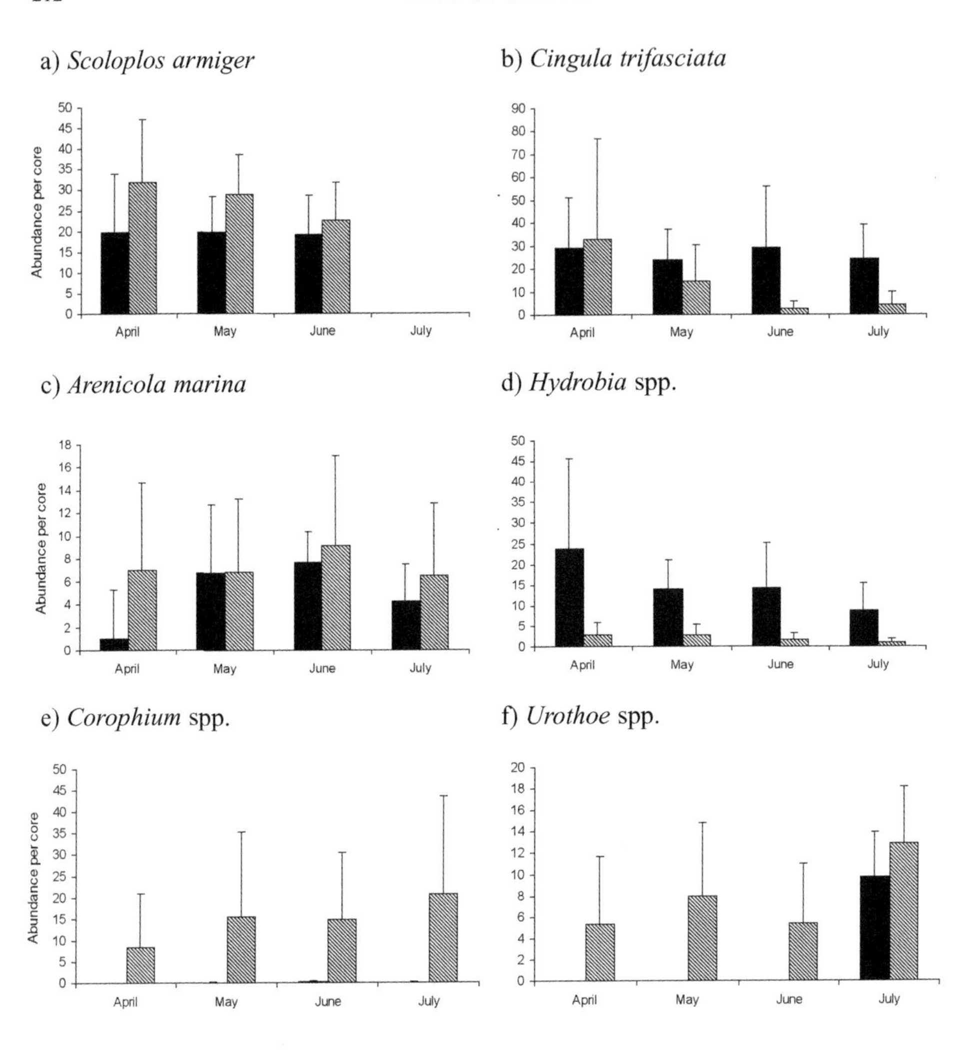

Figure 3 Variation in abundance of key species (Site A: solid bars; Site B: striped bars).

Cingula trifasciata showed relatively little change in abundance at Site A and a reduction at Site B (Figure 3b). These differences were found to be significant (Table 3). *Post hoc* tests indicated that July was significantly different from all other months. In contrast, the abundance of *Arenicola marina* at Site A showed a general increase, whilst at Site B numbers remained relatively constant (Figure 3c). ANOVA indicated that significant differences occurred between site and month (Table 3), with April and June being significantly different from one another.

Hydrobia spp. were more abundant at Site A than Site B, and showed a reduction in abundance at both sites with the onset of cockle dredging (Figure 3d). Significant differences were found between the sites and over the months (Table 3), with April and July being significantly different. In contrast to *Hydrobia* spp., *Corophium* spp. were more abundant at Site B, and were found to increase in abundance during the survey (Figure 3e). At Site A, this species was poorly represented and showed little variation in abundance (Figure 3e). Only site was found to be a significant factor for this species (Table 3). *Urothoe* spp. were also more abundant at Site B compared with Site A (Figure 3f). Increases in the abundance of this species were observed throughout the study period at both sites. This was particularly noticeable for Site A where the species was poorly represented in the first 3 months and showed a large increase in July, almost to the level of abundance observed at Site B (Figure 3f). These variations were found to be statistically significant (Table 3).

Discussion

Effects on the sediment composition

Fishing disturbance can have an indirect effect on benthic communities through the alteration of the substratum (Bradshaw *et al.*, 2002). One of the immediate effects of bottom fishing is to suspend fine sediments into the water column, resulting in a coarsening and destablization of the sedimentary environment (Langton and Robinson, 1990; Messieh *et al.*, 1991; Currie and Parry, 1996). However, Eleftheriou and Robertson (1992) observed rapid resettling of suspended material with little alteration of the sediment size distribution. In the current study, no significant effect on sediment particle size distribution was observed. However, although not investigated in the current study, the layering within the sediment was likely to be affected. This type of modification may lead to changes in the infaunal community through the alteration of elements of the physical environment such as water, oxygen and organic content.

An additional effect of pump-scoop dredging was its aesthetic impact, i.e. the significant scarring of sediment when the tide is out (Figure 4). The scarring effects, however, may only persist for a short period. The rate at which trenches and depressions that result from harvesting activities disappear depends on sediment bed-load transport, suspended sediment load in the water column, exposure to wave action and the harvesting techniques used (Kaiser *et al.*, 2001). Hall *et al.* (1990) observed no detectable effect of suction dredging after 40 days and Hall and Harding (1998) found that trenches made by tractor dredges in the Solway Firth were no longer visible one day after harvesting. Surveys will be required after the close of the cockle fishery in Poole Harbour to assess how long the scarring marks persist.

Effects on the infaunal community

Margalef's and Pielou's indices detected changes in the infaunal community whilst the Shannon Index did not, despite some quite significant reductions in the abundance of individual species. Margalef's Index is a simple total species-abundance ratio, whilst the Shannon Index is based upon proportional abundance of the species. In contrast, Pielou's

Figure 4 Visual impact of pump-scoop dredging in Poole Harbour (source: Linda Parker).

Evenness Index examines the spread of individuals between species. It is a consequence of these differences that the variation in results at the community level was observed. Ghazanshahi *et al.* (1983) and Keough and Quinn (1991) both observed significant changes in abundance of individual species on rocky shores impacted by human activity, however, when the data were used to calculate the Shannon Index, no significant differences were observed. The Shannon Index is unduly influenced by dominant species (Kempton and Taylor, 1976; Pearson and Rosenberg, 1978; Magurran, 1988). In the current study, although the dominant species changed between months, the proportion of dominant species in the community remained relatively constant at approximately 35%, with the second most abundant species contributing approximately 15%. This consistency resulted in very little variation in the Shannon Index.

Messieh *et al.* (1991) and Hall and Harding (1998) reported 40–50% reductions in the abundance of individuals following trawling and tractor dredging activity, respectively. In addition, Hall *et al.* (1990), Brown and Wilson (1997), and Collie *et al.* (2000) also reported significant reductions in infaunal abundance and species composition as a response to suction and other dredging activity. In the current study, however, no significant differences were observed in the infaunal communities between April and May. This indicates that either fishing effort was initially low (scarring marks were, however, observed) or that there was no acute impact of pump-scoop dredging on the benthic community.

Significant differences in the infaunal community were, however, detected 3 months after the cockle fishery opened. By July, overall infaunal abundance had decreased by 42.3% at Site A and by 50.6% at Site B when compared with the closed season for cockle dredging. This change in the benthic community between June and July may be attributable to factors such as a sudden change in temperature, mortality following reproduction or disturbance from a source other than pump-scoop dredging (see below for further discussion). In addition, the reduction in infaunal abundance may be due to an increase in fishery effort. However, little evidence of this was observed in terms of scarring marks. Alternatively, these reductions may be indicative of a chronic effect of pump-scoop dredging. A single pump-scoop dredging event may not have a particularly significant impact on the benthic community. However, over time these disturbances are compounded and eventually the species present succumb to the effect of pump-scoop dredging.

Although these changes may be in response to pump-scoop dredging, it should be noted that other potentially disturbing activities occur at the study sites. These include hand raking for cockles and bait digging. The disturbance caused by hand raking is likely to be relatively similar to that of pump-scoop dredging, although on a much smaller scale. In recent years, bait digging in Poole Harbour has increased dramatically (N. Richardson, pers. comm). Elsewhere this activity has been found to have significant impact on the benthic community (McLusky *et al.*, 1983; Heiligenberg, 1987; Wynberg and Branch, 1994, 1997). Within Poole Harbour, bait digging tends to be done by hand, which has been found to have less of an impact than mechanical harvesting (Heiligenberg, 1987). In addition, because bait digging is a year round activity in the harbour, its impacts are more likely to form a constant background to other disturbances. Further investigation will be required to verify that the disturbance effects observed in the current study were directly related to pump-scoop dredging.

Two processes are likely to play a part in returning the abundance of species in disturbed areas to pre-impact levels: migration by larval and adult infauna, and passive translocation resulting from wind and tide-induced sediment transport (Hall *et al.*, 1990; Hall and Harding, 1998; Ferns *et al.*, 2000). However, the recovery rate of the sediment habitat and its associated fauna is highly variable, depending upon sediment type, local environmental conditions and the type and frequency of harvesting process employed (Kaiser *et al.*, 2001; Piersma *et al.*, 2001). In the current study, it was not possible to examine recovery of the infauna. However, because the sediment is naturally mobile and there is a good local adult population in the surrounding area, it is likely that recovery will be fairly rapid.

One of the original concerns raised regarding pump-scoop dredging in Poole Harbour, was the potential impact on important bird populations, particularly through a reduction in infaunal prey species. Two species commonly cited as important prey are *Arenicola marina* and *Corophium* spp. (e.g. Ferns, 1992). The current research observed no obvious reduction in either of these species as a response to pump-scoop dredging. This suggests that the dredging may not have an obvious detrimental impact on the bird populations through impacts on the infaunal community. However, the noise associated with the dredging activity and the subsequent change to seabed topography may affect bird foraging and other activities.

Summary

Pump-scoop dredge harvesting of cockles does not have an acute impact on the infaunal community of Poole Harbour. However, there may be a chronic effect as the fishery season progresses, causing declines in the abundance of many non-target infaunal species during dredging. Due to the long-term nature of the fishery season, from May to the end of January, it was not possible to determine whether the benthic community recovers from this disturbance. However, since this is not a brand new fishery technique within the harbour, it is likely that recovery is fairly rapid and occurs before the next season begins. Therefore, the environmental impact of these dredging events in the shallow intertidal waters is unlikely to be a factor in their long-term environmental and biological condition. Conflicts with other fishery interests (e.g. bait digging and hand rakers for cockles) and issues related to the biology/ecology of target and non-target species, are elements that need to be addressed before the full impact of the pump-scoop dredging fishery can be assessed.

References

Atkinson, P. W., Clark, N. A., Bell, M. C., Dare, P. J., Clark, J. A. and Ireland, P. L. (2003) Changes in commercially fished shellfish stocks and shorebird populations in the Wash, England. *Biological Conservation,* **114**: 127–141.

Bradshaw, C., Veale, L. O. and Brand, A. R. (2002) The role of scallop-dredge disturbance in long-term changes in Irish Sea benthic communities: a re-analysis of an historical data set. *Journal of Sea Research*, **47**: 161–184.

Broad, G. (1997) An Investigation of the Ecological Effects of Harvesting Cockles (*Cerastoderma edule* L.) by Hand Raking on the Intertidal Benthic Communities in the River Dee Estuary, North Wales. Unpublished MSc Thesis, University of Wales, Bangor.

Brown, B. and Wilson, W. H. Jr. (1997) The role of commercial digging of mudflats as an agent for change of infaunal intertidal populations. *Journal of Experimental Marine Biology and Ecology,* **218**: 49–61.

Clarke, K. (1993) Non-parametric multivariate analyses of changes in community structure. *Australian Journal of Ecology,* **18**: 117–143.

Coffen-Smout, S .S. (1998) Shell strength in the cockle *Cerastoderma edule* L. under simulated fishing impacts. *Fisheries Research,* **38**: 187–191.

Collie, J. S., Hall, S. J., Kaiser, M. J. and Poiner, I. R. (2000) A quantitative analysis of fishing impacts on shelf-sea benthos. *Journal of Animal Ecology,* **66**: 785–799.

Cotter, A. J. R., Walker, P., Coates, P., Cook, W. and Dare, P. J. (1997) Trial of a tractor dredger for cockles in Burry Inlet, South Wales. *ICES Journal of Marine Science,* **54**: 72–83.

Currie, D. R. and Parry, G. D. (1996) Effects of scallop dredging on a soft-sediment community: a large scale experimental study. *Marine Ecology Progress Series,* **134**: 131–150.

Eleftheriou, A and Robertson, M. R. (1992) The effects of experimental scallop dredging on the fauna and physical environment of a shallow sandy community. *Netherlands Journal of Sea Research,* **30**: 289–299.

Ferns, P. (1992) *Bird Life of Coasts and Estuaries*. Cambridge: Cambridge University Press.

Ferns, P. N., Rostron, D. M. and Siman, H.Y., 2000. Effects of mechanical cockle harvesting on intertidal communities. *Journal of Applied Ecology,* **37**: 464–474.

Gaspar, M. B., Leitão, F., Santos, M. N., Sobral, M., Chícharo, L., Chícharo, A. and Monteiro. C. C. (2002) Influence of mesh size and tooth spacing on the proportion of damaged organisms in the catches of the Portuguese clam dredge fishery. *ICES Journal of Marine Science,* **59**: 1228–1236.

Ghazanshahi, J., Huchel, T. D. and Devinny, J. S. (1983) Alteration of Southern California rocky shore ecosystems by public recreational use. *Journal of Environmental Management,* **16**: 37–39.

Hall, S. J., Basford, D. J. and Robertson, M. R. (1990) The impact of hydraulic dredging for razor clams *Ensis* sp. on an infaunal community. *Netherlands Journal of Sea Research,* **27**: 119–125.

Hall, S. J. and Harding, M. J. C. (1998) The effects of mechanical harvesting of cockles on non-target benthic infauna. *Scottish Natural Heritage Research, Survey and Monitoring Report,* No. 86.

Heiligenberg, T. van den (1987) Effects of mechanical and manual harvesting of lugworms *Arenicola marina* L. on the benthic fauna of tidal flats in the Dutch Wadden Sea. *Biological Conservation*, **39**: 165–177.

Hiddink, J. G. (2003) Effects of suction-dredging for cockles on non-target fauna in the Wadden Sea. *Journal of Sea Research,* **50**: 315–323.

Jenkins, J. G. (1991) *The Inshore fishermen of Wales*. Cardiff: University of Wales Press.

Kaiser, M. J., Broad, G. and Hall, S. J. (2001) Disturbance of intertidal soft-sediment benthic communities by cockle hand raking. *Journal of Sea Research,* **45**: 119–130.

Keough, M. J. and Quinn, G. P. (1991) Causality and the choice of measurements for detecting human impacts in marine environments. *Australian Journal of Marine and Freshwater Research*, **42**: 539–554.

Kempton, R. A. and Taylor, L .R. (1976) Models and statistics for species diversity. *Nature*, **262**: 818–820.

Langton, R. W. and Robinson, W. E. (1990) Faunal associations on scallop grounds in the western Gulf of Maine. *Journal of Experimental Marine Biology and Ecology,* **144**: 155–171.

Magurran, A. E. (1988) *Ecological Diversity and its Measurement.* London: Croom Helm.

McLusky, D. S., Andeson, F. E. and Wolfe-Murphy, S. (1983) Distribution and population recovery of *Arenicola marina* and other benthic fauna after bait digging. *Marine Ecological Progress Series,* **11**: 173–179.

Messieh, S. N., Rowell, T. W., Peer, D. L. and Cranford, P. J. (1991) The effects of trawling, dredging and ocean dumping on the eastern Canadian continental shelf seabed. *Continental Shelf Research,* **11**: 1237–1263.

Pearson, T. H. and Rosenberg, R. (1978) Macrobenthic succession in relation to organic enrichment and pollution of the marine environment. *Oceanography and Marine Biology*, **16**: 229–311.

Pickett. G. (1973) The impact of mechanical harvesting on the Thames Estuary cockle fishery. *MAFF Laboratory Leaflet,* No.29.

Piersma, T., Koolhaas, A., Dekinga, A., Beukema, J. J., Dekker, R. and Essink, K. (2001) Long-term indirect effects of mechanical cockle dredging on intertidal bivalve stocks in the Wadden Sea. *Journal of Applied Ecology,* **38**: 976–990.

Rees, E. I. S. (1996) *Environmental Effects of Mechanised Cockle Fisheries: A Review of Research Data.* Bangor: School of Ocean Sciences, University of Wales. Report commissioned by the Marine Environmental Protection Division, Ministry of Agriculture Fisheries and Food..

Rostron D. M. (1995) The effects of mechanised cockle harvesting on the invertebrate fauna of Llanrhician Sands. pp. 111–117. In: *Burry Inlet & Loughor Estuary Symposium: State of the Estuary Report.* Part 2.

Rueda, J. L. and Salas, C. (2003) Seasonal variation of a molluscan assemblage living in a *Caulerpa prolifera* meadow within the inner Bay of Cadiz (SW Spain). *Estuarine, Coastal and Shelf Science,* **57**: 909–918.

Souza, J. R. B. and Gianuca, N. M. (1995) Zonation and seasonal variation of the intertidal macrofauna on a sandy beach of Parana State, Brazil. *Scientia Marina,* **59**: 103–111.

Tuya, F., Perez, J., Medina, L. and Luque, A. (2001) Seasonal variation of the macrofauna from three seasgrass meadows of *Cymodocea nodosa* off Gran Canaria (Central eastern Atlantic Ocean). *Ciencias marinas,* **27**: 223–234.

Underwood, A. T. (1997) *Experiments in Ecology. Their Logical Design and Interpretation Using Analysis of Variance.* Cambridge: Cambridge University Press.

Wynberg, R. P. and Branch, G. M. (1994) Disturbance associated with bait collection for sandprawns (*Callianassa kraussi*) and mudprawn (*Upogebia africana*) longterm effects on the biota of intertidal sandflats. *Journal of Marine Research,* **52**: 523–558.

Wynberg, R. P. and Branch, G. M. (1997) Trampling associated with bait collection for sand prawns *Callianassa kraussi* Stebbing effects on the biota of an intertidal sandflat. *Environmental Conservation,* **24**: 139–148.

The Ecology of Poole Harbour
John Humphreys and Vincent May (editors)

18. Water Quality and Pollution Monitoring in Poole Harbour

Julian Wardlaw

Environment Agency, Rivers House, Sunrise Business Park, Upper Shaftesbury Road, Blandford Forum, Dorset

The waters of Poole Harbour currently achieve the statutory water quality standards laid down by the European Union. Water pollution from human and industrial activity in the last century has caused lasting damage to the ecological balance in Poole Harbour. Toxic pollutants have been concentrated in sediments, particularly in Holes Bay where circulation and flushing is restricted, and will take many years to recover. While many of the historically polluting point source discharges to the harbour have been eliminated or improved in quality, diffuse metal, bacteriological and nutrient sources remain a threat to wildlife.

Introduction

Poole Harbour is an estuary of nearly 4000 ha with an unusual double-high tide and micro-tidal regime. The narrow opening at the eastern end ensures that only 22% of the harbour water at neap tides, and 45% on spring tides, is returned to Poole Bay. This poor flushing ensures that, particularly in inner zones such as Holes Bay which also has a narrow entrance, pollutants are easily trapped. Fine intertidal muds offer poor dispersion in the bay for pollutants, especially metals and organic chemicals.

The wide diversity of habitats and limited human access allows many species to thrive, giving rise to high conservation status as an SSSI (Site of Special Scientific Interest), RAMSAR Wetland Site and Special Protection Area (SPA).

Legislation

The first attempts to introduce legal controls over aqueous discharges were made in England in 1936, mainly to combat increasing public health issues from sewage. More specific anti-pollution legislation was enacted in 1951 for rivers and 1960 for tidal waters, with controls over trade effluent discharges to foul sewers tightened in 1937 and 1974. The current powers of the Water Resources Act 1991 give strong control over consented discharges and illegal pollution.

Effective regulation is a combination of statutory powers and commitment, which was perhaps lacking until the late twentieth century. EU Directives currently govern the majority of harbour monitoring, providing statutory European standards for Bathing Waters, Dangerous Substances, Urban Wastewater, Shellfish Hygiene, Shellfish Waters, Nitrates and water basin management.

Future development must be mindful of the Habitats Directive, the Birds Directive, the Water Framework Directive and the internationally important conservation status of Poole Harbour for birds and wildlife.

Human pressures on Poole Harbour

Socio-economic conditions have dominated harbour uses, as can be seen today in the general split between development and commerce in Poole town and natural protected areas in the southern harbour. It was not always this way: early development focused on exploiting clay and mineral deposits to the south of the harbour and on Brownsea Island. Mineral extraction remains, but is largely hidden from view: oil production from the southern side of the harbour in the biggest onshore oilfield in western Europe is a shining example of how industrial development need not damage the environment if the will exists.

Industry in the twentieth century has left a deep scar on the northern shores of the harbour. With the benefit of hindsight, the choice of location for chemical manufacture, metal plating and sewage treatment could probably not have been much worse. Holes Bay in particular has suffered: the discharge of treated sewage effluent was first consented in the 1950s, no doubt preceded by untreated local discharges. This domestic sewage discharge was combined with significant chemical pollutant loads from illegal trade effluent (especially heavy metals) with poor application of the legislation, until as recently as the late 1970s. Although now treated to extremely high standards (both chemically and disinfected), nutrients in the discharge remain a threat to wildlife, currently encouraging spectacular growths of algae. Many threats remain despite removal of some industry: land has been reclaimed with landfill, and housing and light industry generate their own pollution pressures in the form of poor management, illegal drainage connections and contaminant runoff.

Significant pollutants

Heavy metals

Metal contamination within the harbour is largely restricted to Holes Bay, affecting bioaccumulation in molluscs and crustaceans, species composition, larval fish and birds. Decades of toxic metal discharges have occurred, passing through the foul and surface water sewers, to a low-energy intertidal zone with restricted dispersion, causing concentration of contaminated particles throughout the bay. The sediments have accumulated these toxic elements, including cadmium and mercury, and locally silver, copper, zinc and selenium.

The fate of these is unknown and natural breakdown is slow, but it has been suggested that these metals may be labile, and could re-mobilize when disturbed. The use of sacrificial zinc anodes on sheet piling and commercial and recreational craft has caused local effects throughout the harbour and is currently being studied.

TBT and organic pollutants

Used widely for anti-fouling, tributyltin (TBT) was found, in the 1980s, to have severely damaging effects upon marine organisms, causing changes such as deformities, shell thickening and death. High concentrations were found in most of the northern shoreline in the water and in sediments. TBT was banned in 1987 for craft smaller than 25 m, but continues to affect harbour mollusc populations on northern shores. There has been little recorded pollution by organic chemicals, but local pollution by wood preservative at Holton Heath in the late 1980s introduced toxic pentachlorophenol, dieldrin and lindane to the local environment.

Nutrients

High nutrient levels in harbour waters have led to Poole Harbour being designated in 2002 as a Sensitive Area (Eutrophic) and Polluted Waters (Eutrophic) under the Urban Wastewater and Nitrate Directives respectively, and the river catchment area being designated a Nitrate Vulnerable Zone (NVZ) (Langston *et al.* (2003)). The main sources of elevated nitrate and phosphate in the harbour are Poole Sewage Treatment Works and the riverine inputs from the Rivers Frome and Piddle containing diffuse pollutants. High nutrient inputs have caused macro-algal and diatom blooms, and may be responsible for shellfish mortalities between 1995 and 1997 (including partial closure of the shellfishery).

Some 50,000 m^3 of treated sewage effluent discharge daily from Poole Sewage Treatment Works, contributing approximately 1500 kg of nitrogen per day. It also contributes some 80% of the input of phosphate to the harbour. Treatment improvements before 2010 should address this nutrient load, but it may be many years before these reductions are observed in positive changes to flora and fauna. Contributing rivers derive most nitrate from agricultural sources, and the Rivers Frome, Piddle, Corfe and Sherford are designated as NVZs. New mandatory restrictions in slurry storage during winter and fertilizer applications will reduce nitrogen input.

Monitoring

Routine harbour water quality monitoring is dominated by the statutory requirements of EU Directives. In 2002, no monitoring site failed to meet the EU standards, reflecting the improving waters of the harbour. This monitoring includes water, shellfish and some limited sediment analysis. Elevated bacteriological results have caused local downgrading of the shellfishery, and are currently under investigation.

Major harbour inputs (Rivers Frome, Piddle, Corfe and Sherford) are routinely sampled as well as consented discharges. Combined storm overflows (CSOs) exist and discharge sporadically, but none are known to cause environmental damage.

The legacy of previous pollution remains in the sediments, primarily in Holes Bay. These are not routinely examined (except for limited sites under the Dangerous Substances Directive), other than when development is proposed, but some exceptional surveys have been undertaken, indicating slow improvement.

It is clear that monitoring could be improved to provide a more comprehensive picture of the current and future states of the harbour, but unless significant funding is forthcoming, this is unlikely to happen.

Future challenges

The legacy of historic pollution will remain in the harbour for many years, particularly in Holes Bay, but significant measures have been taken to prevent current discharges from causing further damage. While direct and diffuse discharges remain, nutrient reduction remains the major challenge. Some effort is also required to examine the threat of zinc, and to determine whether all elevated levels of bacteria have been addressed. Future development must be mindful of previous damage, and of the need to prevent pollution in this very sensitive environment.

References

Langston, W. J., Chesman, B. S., Burt, G. R., Hawkins, S. J., Readman, J. and Worsfold, P. (2003) *Site Characterisation of the South West European Marine Sites – Poole Harbour SPA*. A study carried out by Plymouth Marine Science Partnership on behalf of the Environment Agency and English Nature.

The Ecology of Poole Harbour
John Humphreys and Vincent May (editors)

19. Sediment Quality and Benthic Invertebrates in Holes Bay

Fiona Bowles[1] and Paul English[2]

[1]Wessex Water, Claverton Down Road, Bath, BA2 7WW

[2]Emu Ltd, 1 Mill Court, The Sawmills, Durley, Southampton S032 2EJ

This chapter presents summary comparisons between two sets of physical, chemical and biological data collected in 1991 by the Environment Agency and in 2002 by the Centre for Ecology and Hydrology. Changes in the invertebrate data have been assessed in the context of natural variation and in relation to the present physical and chemical conditions.

Introduction

Poole Sewage Treatment Works serves a population equivalent of over 150,000 with inputs from domestic, tourism and local industry sources. The works is located on Cabot Lane and discharges to Holes Bay at SZ 00710 93560. The works, which has been present since 1922, has two streams. The western stream was developed in 1957–61 and the eastern stream was added in 1969–74. Both are activated sludge. In 1994, a new inlet works was added and in 1996, a biological aerated filter replaced the old eastern stream. Under the Urban Wastewater Treatment Directive, Poole Works was required to have ultra violet disinfection added by March 2003 to ensure bathing water and shellfish water quality for Poole Harbour and its beaches.

In order to ensure effective disinfection, additional settlement of suspended solids in the treated effluent was required. Ferric sulphate is the proposed flocculent, although the discharge consent will permit the use of an aluminium salt if the UV performance deteriorates. As there is no current environmental quality standard for aluminium in the marine environment, the freshwater standard of 1 mg l^{-1} aluminium was assumed and English Nature requested that a baseline survey of the biota and metal levels be established in Holes Bay. Wessex Water, part of the Poole Harbour Steering Group, therefore, contracted the Centre for Ecology and Hydrology to extend their survey of prey distribution in Poole Harbour to collect:

- additional invertebrate samples in Holes Bay
- sediment samples for analysis of particle size, carbon, nitrogen and metal parameters
- samples of a numerous bivalve, the Edible Cockle *Cerastoderma edule*, for tissue analysis of metal parameters.

In order to gain comparative data, the sample sites chosen were among those that had been previously sampled by the Environment Agency in 1991 and 1996 and are illustrated in Figure 1. Richard Caldow and colleagues from the Centre for Ecology and Hydrology conducted the survey. Chemical samples were analysed by the Environment Agency (Starcross Laboratory). Paul English of Emu Ltd, Southampton, was commissioned to compare the available ecological data for 1991 and 2002 and to assess current sediment quality with respect to the potential effects on the local bird feeding/sediment invertebrate interest.

Methodology

Ten sites were selected from 46 previously sampled locations (shown in Figure 1) and located with hand-held GPS. The methodology for benthic invertebrate and sediment samples was as for the Poole Harbour bird prey survey (see chapter 7). In addition, at each of the sites, 500 ml of the surface sediment (maximum depth 1 cm and avoiding black anoxic material if possible) was collected for metal (As, Cd, Cr, Cu, Pb, Ni, Zn, Fe, Hg and Al), carbon and nitrogen analysis, with a duplicate sample being collected at five of the sites. In addition, at six sites (1, 3, 4, 5, 17 and 28), 30 cockles were caught in hand-held dredging nets and depurated for 72 hours in clean seawater prior to being frozen and sent for tissue analysis.

Results and discussion

Particle size analysis and organic ratios

Fine sediment, except at the outermost site (46), dominated the sediment. Since 1991, a slight coarsening of the sediments is evident with all stations showing an increased fine sand component, particularly the lower mudflat stations (Stations 1, 17, 28 and 34), where some of the greatest changes in cirratulid numbers have also occurred. The bay has become less muddy.

Ratios of organic carbon to organic nitrogen of 10:1 to 15:1 and 11:1 to 16:1 were found in 1991 and 2002, respectively. Values of 7:1–12:1 are typical of shallow marine sediments but estuarine sediment usually exhibits higher ratios, as a result of high organic carbon inputs from fringing plant communities or riverine inputs (Murray *et al.*, 1980). Sewage influence would lead to a lower ratio due to the input of nitrogenous compounds. The carbon:nitrogen ratios are typical of estuarine sediments and indicate a dominant, natural terrestrial source with no significant influences from the outfall.

Metals

The sediment comparison is shown in Table 1. Sediment concentrations of iron and arsenic have significantly decreased ($P<0.05$) over the 11 year period between 1991 and 2002, whilst aluminium levels have significantly increased ($P<0.05$). All sampling stations returned raised aluminium levels with the most notable at Station 6, adjacent to the outfall

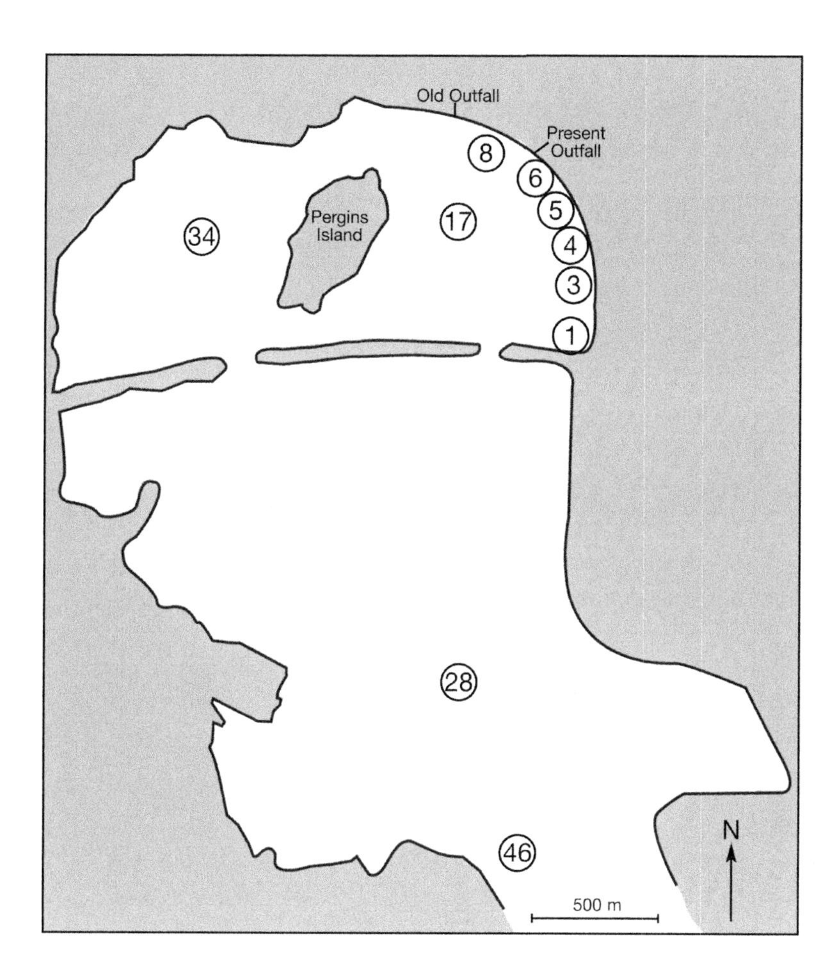

Figure 1 Holes Bay showing the sampling stations. (Station numbers relate to original surveys in the 1990s

(+28.2%), Station 28 downstream (+32.6%), and Station 34 (+30.8%), on the western side of the bay. The metal levels did not appear to relate to the invertebrate populations.

When corrected to 1% organic carbon, the present metal levels are all below the UK proposed tentative action levels (CEFAS, 1997) and the US-developed thresholds (Zarba, 1989), where available (Table 2). There is no specific toxicity data available relating to aluminium and the fauna distribution did not show any significant adverse effect from the sediment quality.

The cockle tissue data form a baseline for any future changes and no comparison with historic levels was possible.

Table 1 Statistical comparison between sediment chemical determinands

Station	Fe 1991 ($mg\ kg^{-1}$)	Fe 2002 ($mg\ kg^{-1}$)	Al 1991 ($mg\ kg^{-1}$)	Al 2002 ($mg\ kg^{-1}$)	As 1991 ($mg\ kg^{-1}$)	As 2002 ($mg\ kg^{-1}$)	Org C 1991 (%)	Org C 2002 (%)	Org N 1991 (%)	Org N 2002 (%)
1	32300	23900	9410	11400	14.60	11.00	3.08	2.95	0.30	0.22
3	39300	25100	13400	14100	14.50	11.00	3.48	3.62	0.35	0.27
4	35200	24800	11500	12950	10.20	11.50	3.39	3.71	0.33	0.23
5	35100	23700	10700	12100	14.60	11.00	3.61	3.88	0.35	0.28
6	28300	27150	8940	12450	10.30	12.50	3.76	4.94	0.35	0.41
8	37600	27150	12100	12550	15.80	12.00	3.92	5.01	0.33	0.41
17	36000	20200	10800	10600	10.10	9.00	3.22	3.73	0.32	0.29
28	21800	20000	6620	9820	12.90	9.00	1.62	3.68	0.11	0.35
34	30200	24350	8340	12050	16.10	12.00	2.26	3.73	0.20	0.37
46	27600	19150	10400	11400	15.50	11.00	2.59	1.05	0.24	0.07
average	**32340**	**23550**	**10221**	**11942**	**13.46**	**11.00**	**3.09**	**3.63**	**0.29**	**0.29**
var	**28862667**	**8206111**	**3852588**	**1469862**	**5.83**	**1.39**	**0.53**	**1.21**	**0.01**	**0.01**
	Log	Log	Log	Log						
	4.51	4.38	3.97	4.06						
	4.59	4.40	4.13	4.15						
	4.55	4.39	4.06	4.11						
	4.55	4.37	4.03	4.08						
	4.45	4.43	3.95	4.10						
	4.58	4.43	4.08	4.10						
	4.56	4.31	4.03	4.03						
	4.34	4.30	3.82	3.99						
	4.48	4.39	3.92	4.08						
	4.44	4.28	4.02	4.06						
average	**4.50**	**4.37**	**4.00**	**4.08**						
var	**0.01**	**0.003**	**0.01**	**0.00**						
f test	**0.30**		**0.05**		**0.04**		**0.24**		**0.46**	
t test	***P*<0.05**		***P*<0.05**		***P*<0.05**		***P*>0.05**		***P*>0.05**	

Invertebrate populations

The benthic populations were consistent with the Joint Nature Conservancy Committee (JNCC) biotope classifications LMU.HedOl on upper shore areas and LMU.HedStr on mid to lower intertidal zones and are typical of sheltered sandy mud – mud and reduced salinity conditions (Connor *et al.*, 1997).

Over 30 species of invertebrates were recovered in 2002, the dominant species being the ragworm *Hediste diversicolor* with *Hydrobia ulvae*, *Cirratulus fliformis* and *Malacocerus fugilinosus*. The 1991 and 2002 top ranking species are compared in Table 3. Ragworms appear to have reduced throughout Holes Bay, while tubificid worms have declined in abundance particularly at Stations 5 and 6 adjacent to the sewage outfall. The spionid worm *Streblospio shrubsolii* has been replaced by *Malacocerus fuliginosus* throughout and by cirritulid worms at the lower mudflat sites. This change correlates with changes in sediment but it is unclear whether this caused it.

Mud Shrimp *Corophium volutator* abundance declined from an average of 771 m^{-2} to 25 m^{-2} in 2002. However, this may be a reflection of its patchy distribution or of localized salinity changes, for example, at Station 1. It is still present at Station 34 on the western side (212 m^{-2}).

Multivariate analysis of the biological data suggests that there were differences in the abundance of characteristic mudflat species, including ragworms, tubificid worms and total spionid worms between 1991 and 2002. However, these differences may be within the expected natural variation or within the variation resulting from the methodological differences, e.g.

- changes over an 11 year period due to natural climatic variations or anthropogenic pressures, such as sediment disturbance
- seasonal difference in the timings of the 1991 (June) and 2002 (September) surveys
- different biological sampling methods employed in 1991 and 2002.

In conclusion, Holes Bay remains biologically, physically and chemically within the range of normal estuarine conditions. No significant adverse effects on the infauna could be attributed to the metal levels. Continued coarsening of the sediments may be having a more pronounced effect on the invertebrate populations in the bay and hence the bird prey availability.

Analysis of the treated effluent annual average shows an increase from 0.04 mg l^{-1} in 1996 to 0.17 mg l^{-1} aluminium. Its source is, however, the incoming sewage itself rather than any dosing. Annual averages for iron levels have varied without an obvious trend from 0.34 g l^{-1} to 0.84 g l^{-1} since 1995. The consented use of iron or aluminium flocculent from 2003 is considered unlikely to impact on the infauna because:

Table 2 Comparison of sediment quality with metals quality guidelines

Determinand	MAFF (DEFRA) action levels (mg kg^{-1})	US-developed threshold (mg kg^{-1})	Holes Bay 2002 mean concentrations (mg kg^{-1}) (corrected to 1% OC)
Aluminium	N/d	N/d	3295.9
Iron	N/d	N/d	6499.5
Arsenic	8	33	3.0
Lead	40	132	19.5
Nickel	100	20	5.9
Copper	40	136	14.2
Chromium	100	25	10.8
Zinc	200	760	58.8
Cadmium	2	31	0.3
Manganese	N/d	N/d	45.0
Mercury	0.4	0.8	0.20

N/d=no data.

Table 3 Top ranking macro-invertebrates for 1991 and 2002 sampling occasions

1991			2002		
Taxa	Mean abundance (m^2)	No. stations present (10)*	Taxa	Mean abundance (m^2)	No. stations present (10)*
Hediste diversicolor	2843	10	*Hediste diversicolor*	1630	8
Tubificidae	1218	9	*Hydrobia ulvae*	1427	10
Corophium volutator	771	8	Cirratulidae	1317	6
Total spionids	433	10	Total spionids	1151	9
Streblospio shrubsolii	430	10	*Malacocerus fuliginosus*	1130	9
Abra tenuis	408	9	Tubificidae	614	8
Hydrobia ulvae	154	8	*Cyathura carinata*	445	9
Cyathura carinata	30	3	*Abra tenuis*	398	7
Actiniaria spp.	15	1	*Actiniaria* spp.	292	3
Terebellida	15	1	*Terebellida*	123	2

*Number of sample stations.

- the incoming metals are largely associated with sludge in sewage treatment so the additional metal flocculent will increase metal removal from the effluent, such that discharge levels should remain less than 50% of the incoming effluent (absolute consent limits are 3 mg l^{-1} of iron or 5 mg l^{-1} of aluminium)
- the species currently predominant in the upper mudflat, close to the sewage discharge, are ragworms, tubificids and spionids, which are tolerant to disturbance and pollution (Pearson and Rosenberg, 1978).

Acknowledgements

This chapter reflects the views of the authors and is not a necessarily the view of Wessex Water Services Ltd. Thanks are owed to Nicole Price of the Environment Agency, Blandford, for arranging depuration of bivalve samples prior to analysis.

References

CEFAS (1997) Marine Pollution Monitoring Group. Final Reports of the Metals Task Team and the Organics Task Team. *Aquatic Environmental Monitoring Report*. Lowestoft: CEFAS.

Connor, D. W., Brazier, D. P., Hill, T. O. and Northern, K. O. (1997) *Marine Nature Conservation Review: Marine Biotope Classification for Britain and Ireland* Volume 1. *Littoral Biotopes*. Version 97.06 JNCC Report, No 229. Peterborough: Joint Nature Conservancy Committee.

English, P. (2003) Holes Bay Intertidal Ecological Data Review and Assessment. Report for Wessex Water. November 2003 (in draft).

Murray, L. A., Norton, M. G., Nunny, R. S. and Rolf, M. S. (1980) The field assessment of dumping wastes at sea: 6. The disposal of sewage sludge and industrial waste off the River Humber. *Fisheries Research Technical Report*, No. 55. Lowestoft: Ministry of Agriculture, Fisheries and Food.

Pearson, T. H. and Rosenberg, R. (1978) Macroinvertebrate succession in relation to organic enrichment and pollution of the marine environment. *Oceanography and Marine Biology Annual Review*, **16**: 229–311.

Zarba, C. (1989) *National Perspective on Sediment Quality*. Committee on Contaminated Sediments – Assessment and Remediation (ed.). Washington DC: National Academy Press.

20. Macroalgal Mat Development and Associated Changes in Infaunal Biodiversity

Eunice Pinn and Martin Jones

School of Conservation Sciences, Bournemouth University, Talbot Campus, Fern Barrow, Poole, Dorset BH12 5BB

Blooms of macroalgal matting, comprising opportunistic species such as *Ulva lactuca*, are becoming increasingly common in Poole Harbour. A hostile environment is usually created in the sediment below a dense algal mat, influencing the invertebrate faunal assemblage. This preliminary study was conducted over a 6 month period during which a dense mat of *U. lactuca* developed and subsequently dispersed in Holes Bay. The algal mat was found to have a significant negative impact on species richness, abundance and biomass of the infauna. The results are discussed in relation to impacts on the ecosystem as a whole.

Introduction

Although blooms of ephemeral green algae are a natural component of estuarine habitats (Everett, 1991), they are becoming increasingly prevalent around the world (Morand and Briand, 1996; Pihl *et al.*, 1999; Viaroli *et al.*, 2001). These exceptional blooms are thought to be indicators of anthropogenically induced eutrophication at the sediment/water interface (Everett, 1991; Fletcher, 1996). The presence of these mats can lead to major changes in the biogeochemical cycles (Morand and Briand, 1996; Valiela *et al.*, 1997), which can result in modification of food chains, faunal community structure and ecosystem processes (Valiela *et al.*, 1997; Raffaelli *et al.*, 1998).

At the base of the mat, an anoxic gradient gradually develops due to decomposition of the algae (Bolam *et al.*, 2000). In addition, the water within the mat can become super-saturated (Krause-Jensen *et al.*, 1999), which leads to severe diurnal fluctuations in oxygen (D'Avanzo and Kremer, 1994). Bacterial decomposition has been demonstrated to increase dramatically within both the sediment and the water column (Nedergaard *et al.*, 2002), leading to hypoxia and anoxia. This can be prolonged, resulting in the accumulation of sulphides due to the activity of sulphate-reducing bacteria (Viaroli *et al.*, 2001). It is these biochemical changes that result in modifications to the macrofaunal community and, consequently, effects higher up the food chain

In recent years, macroalgal mats of *Ulva lactuca* have become increasingly common in Poole Harbour. Concerns have been expressed by the RSPB and English Nature regarding the impact of the annual mats on the wildfowl using the harbour. The aim of this study was to assess this impact, in part, by investigating the effect of the macroalgal bloom on the invertebrate fauna of the harbour.

Methods and materials

The study area was sited within Holes Bay, an enclosed bay of Poole Harbour with limited tidal flushing, and conducted from June to November 2002 at a permanently marked site measuring 50 m x 50 m. Due to the nature of the site, it was not possible to have a control position where no algal mat development occurred.

At monthly intervals, 30 0.25 m^2 randomly placed quadrats were used to determine the percentage coverage of the algae; 30 75 mm diameter sediment cores were also extracted on a monthly basis, to a depth of 150 mm. The sediment obtained was washed through a 0.5 mm sieve and the macrofaunal obtained recorded to species level.

Analysis of variance (ANOVA) was undertaken for species richness (i.e. number of species present per core), the infaunal abundance (i.e. total number of individuals per core) and biomass (i.e. wet weight (g) of infauna per core including shells of live molluscs) data. Prior to analysis, data were tested for homogeneity of variances using Levene's test. Data transformation, however, did not remove the heterogeneity. Underwood (1997) reported that for large balanced data sets, violations in the assumption of homogeneity and normality were unlikely to affect the F ratio. It was, therefore, decided to undertake the ANOVA using the non-transformed data, but with a more conservative probability of 0.01 (Connell, 2001).

Results

Within 2 months of the first visit in June, a dense algal mat had developed. Mean coverage increased to a maximum of 91.0% in August (Figure 1). Thereafter, mean coverage declined, reducing to 3.8% by November when the survey finished.

A total of 15 infaunal species were identified from Holes Bay (Table 1). Infaunal richness was generally very low on a monthly basis, with a maximum of 1.5 ± 0.9 per core observed in June. Species richness then declined to 0.6 ± 0.7 per core in August as the mat developed (Figure 2a). Thereafter, species richness remained at a reduced level, with the lowest value being recorded in November (0.6 ± 0.6 per core). Using ANOVA, these differences were found to be statistically significant ($P<0.001$, f [5, 174] = 7.212).

Infaunal abundance declined steadily from a maximum of 3.7 ± 2.2 per core in June to a minimum of 0.9 ± 1.2 per core in September, with the lowest value recorded in November (0.8 ± 0.9 per core) (Figure 2b). ANOVA revealed these differences to be significant ($P<0.001$, f [5, 174] = 14.229). Infaunal biomass peaked in July (0.55 ± 0.74 g per core), whilst August had the lowest biomass (0.15 ± 0.37 g per core). In October and November, infaunal biomass appeared to increase to values close to those observed

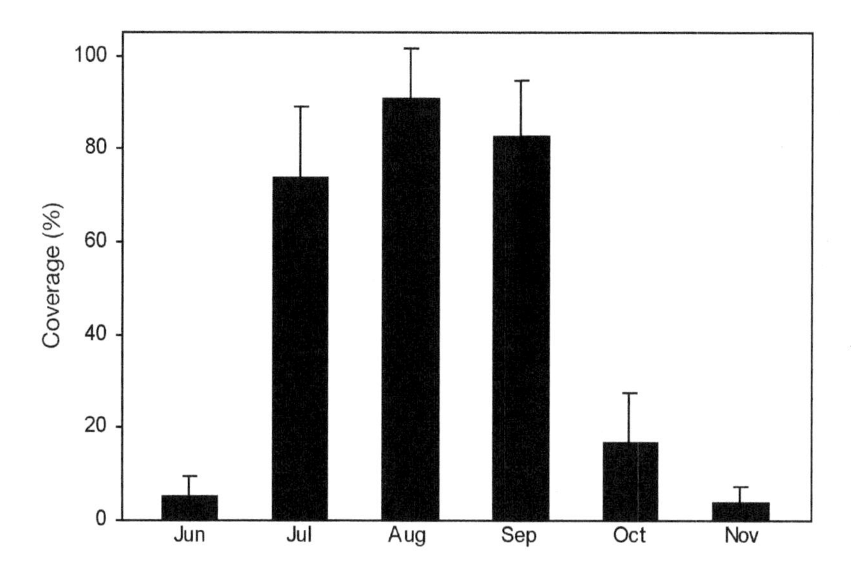

Figure 1 Development of the macroalgal mat.

at the start of the study (Figure 2c). These variations were found to be statistically significant ($P < 0.01$, f [5, 174] = 3.143).

Discussion

In this preliminary 6 month study, marked changes were observed in the invertebrate community of Holes Bay. It should be noted that due to the lack of a suitable control site the changes observed cannot be causally linked to mat development. However, over the spring/summer period, when this study was conducted, the infaunal community would normally be expected to have the highest levels of species diversity and abundance of any time of year (Souza and Gianuca, 1995; Tuya *et al.*, 2001; Rueda and Salas, 2003). It is likely, therefore, that the marked changes observed in the community are associated with the development of the macroalgal mat.

As the mat developed, there was an initial increase in infaunal diversity and abundance. However, this rapidly declined as the mat became more dense. Similar observations associated with the development of macroalgal blooms have been made by Lopes *et al.* (2000) and Bolam *et al.* (2000). The changes occurring in the benthic community in relation to macroalgal blooms are extremely complex (Hull, 1988; Raffaelli *et al.*, 1998). The effects of macroalgal blooms are often similar to those resulting from organic enrichment, including an increase in opportunistic species such as *Capitella capitata* (Bolam *et al.*, 2000; Lopes *et al.*, 2000).

a) Species richness

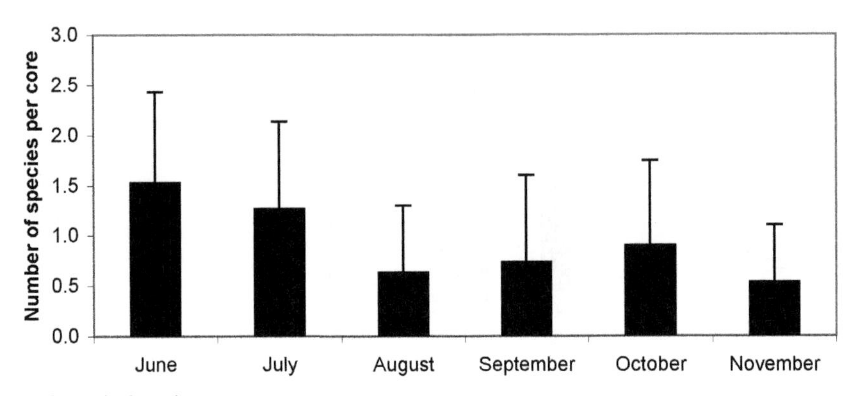

b) Infaunal abundance

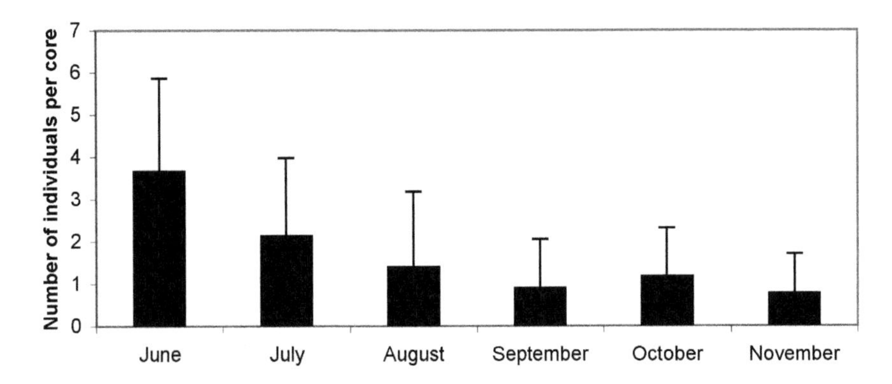

c) Infaunal biomass

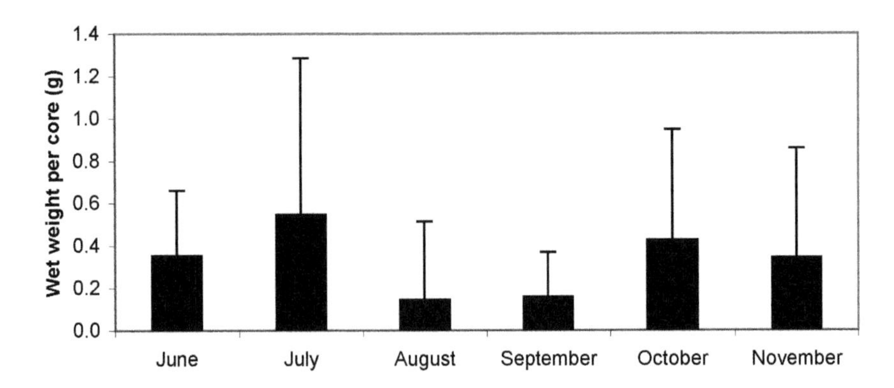

Figure 2 Variation in infaunal community.

Table 1 The infaunal community

	June	July	August	September	October	November
Hediste diversicolor	C	C	C	C	C	C
Nereis zonata	C					
Nereis pelagica	C	C		C		
Perineris cultifera	C			C		
Arenicola marina						C
Nephtys caeca					C	
Capitella capitata					C	
Hydrobia ulvae	C	C	C	C		
Tapes descussatus	C		C	C	C	
Lutraria lutraria	C	C				
Cerastoderma edule		C				
Gari costulata		C				
Venerupis senegalensis						C
Carcinus maenus		C				
Peachia cylindrica	C	C	C	C	C	

C = present in cores.

Hediste diversicolor is a typical estuarine species and the dominant member of the infauna observed in the current study. Conflicting reports on the impact of algal mats on this species have been published, ranging from increasing abundance (Norkko and Bonsdorff, 1996; Österling and Pihl, 2001), through initial increases then declines (Lopes *et al.*, 2000), to steady declines (Lewis *et al.*, 2003). In the current study, the abundance of this polychaete declined steadily, whilst biomass measurements did not decline initially. This may indicate that initially *H. diversicolor* gained from the impact of the mat on other species, but subsequently succumbed itself to the environmental impacts of the algal bloom.

Bivalve molluscs are another important component of estuarine communities. Everett (1994) and Österling and Pihl (2001) found that bivalves such as *Macoma balthica* and *Cerastoderma* spp. decreased in abundance under macroalgal mats. In contrast, Hull (1988) and Bolam *et al.* (2000) reported greater numbers of bivalves. In the current study, although mollusc abundance initially increased, latterly it declined with almost all bivalve species being lost from the system by November. Bolam *et al.* (2000) proposed that these differences relate to algal biomass, with bivalve numbers only declining at higher algal densities.

Any long-term altering of the benthic community could have a cascading effect on ecosystem function (Franz and Friedman, 2002). A shift to ephemeral green algae from other habitat types will result in dramatic alterations in the invertebrate community and complexity of the ecosystem. In some areas, this has already had an effect higher up the food chain (e.g. Raffaelli, 1999; Lewis *et al.*, 2003). It will be necessary to undertake further, more detailed surveys in Holes Bay before the effect of the annual macroalgal bloom can be assessed. It is likely, however, that the bloom will impact on the important wildfowl populations of the area.

References

Bolam, S. G., Fernandes, T. F., Read, P. and Raffaelli, D. (2000) Effects of macroalgal mats on intertidal sandflats: an experimental study. *Journal of Experimental Marine Biology and Ecology*, **249**: 123–137.

Connell, S. D. (2001) Urban structures as marine habitats: an experimental comparison of the composition and abundance of subtidal epibiota among pilings, pontoons and rocky reefs. *Marine Environmental Research,* **52**: 115–125.

D'Avanzo, C. and Kremer, N. J. (1994) Diel oxygen dynamics and anoxic events in an eutrophic estuary of Waquoit Bay, Massachusetts. *Estuaries*, **17**: 131–139.

Everett, R. A. (1991) Intertidal distribution of infauna in a central California lagoon: the role of seasonal blooms of macroalgae. *Journal of Experimental Marine Biology and Ecology*, **150**: 223–247.

Everett, R. A. (1994) Macroalgae in marine soft-sediment communities: effects on benthic faunal assemblages. *Journal of Experimental Marine Biology and Ecology*, **175**: 253–274.

Fletcher, R. L. (1996) The occurrence of green tides – a review. pp. 7–43. In: *Marine Benthic Vegetation: Recent Changes and the Effects of Eutrophication.* Schramm, W. and Nienhuis, P. K. (eds). Berlin: Springer.

Franz, D. R. and Friedman, I. (2002) Effects of a macroalgal mat (*Ulva lactuca*) on estuarine sand flat copepods: an experimental study. *Journal of Experimental Marine Biology and Ecology*, **271**: 209–226.

Hull, S. C. (1988) The Growth of Macroalgal Mats on the Ythan Estuary, with Respect to their Effects on Invertebrate Abundance. Unpublished PhD thesis, University of Aberdeen.

Krause-Jensen, D., Christensen, P. and Rysgaard, S. (1999) Oxygen and nutrient dynamics within mats of the filamentous macroalgae *Chaetomorpha linum*. *Estuaries*, **22**: 31–38.

Lewis, L. J., Davenport, J. and Kelly, T. C. (2003) Responses of benthic invertebrates and their avian predators to the experimental removal of macroalgal mats. *Journal of the Marine Biological Association of the UK*, **83**: 31–36.

Lopes, R. J., Pardal, M. A. and Marques, J. C. (2000) Impact of macroalgal blooms and wader predation on intertidal macroinvertebrates: experimental evidence from the Mondego Estuary (Portugal). *Journal of Experimental Marine Biology and Ecology*, **249**: 165–179.

Morand, P. and Briand, X. (1996) Excessive growth of macroalgae: a symptom of environmental disturbance. *Botanica Marina*, **39**: 491–516.

Nedergaard, R. I., Risgaard-Petersen, N. and Finser, K. (2002) The importance of sulfate reduction associated with *Ulva lactuca* thalli during decomposition: a mesocosm experiment. *Journal of Experimental Marine Biology and Ecology*, **275**: 15–29.

Norkko, A. and Bonsdorff, E. (1996) Population responses of coastal zoobenthos to stress induced by drifting algal mats. *Marine Ecology Progress Series*, **140**: 141–151.

Österling, M. and Pihl, L. (2001) Effects of filamentous green algal mats on benthic macrofaunal functional feeding groups. *Journal of Experimental Marine Biology and Ecology*, **263**: 159–183.

Pihl, L., Svenson, A., Moksnes, P.-O. and Wennhage, H. (1999) Distribution of green algal mats throughout shallow soft bottoms of the Swedish Skagerrak archipelago in relation to nutrient sources and wave exposure. *Journal of Sea Research*, **41**: 281–294.

Raffaelli, D. (1999) Nutrient enrichment and trophic organisation in an estuarine food web. *Acta Oecologica*, **20**: 449–461.

Raffaelli, D. G., Raven, J. R. and Poole, L. (1998) Ecological impact of green macroalgal blooms. *Annual Review of Marine Biology and Oceanography*, **36**: 97–125.

Rueda, J. L. and Salas, C. (2003) Seasonal variation of a molluscan assemblage living in a *Caulerpa prolifera* meadow within the inner Bay of Cadiz (SW Spain). *Estuarine, Coastal and Shelf Science*, **57**: 909–918.

Souza, J. R. B. and Gianuca, N. M. (1995) Zonation and seasonal variation of the intertidal macrofauna on a sandy beach of Parana State, Brazil. *Scientia Marina*, **59**: 103–111.

Tuya, F., Perez, J., Medina, L. and Luque, A. (2001) Seasonal variation of the macrofauna from three seagrass meadows of *Cymodocea nodosa* off Gran Canaria (Central eastern Atlantic Ocean). *Ciencias Marinas*, **27**: 223–234.

Underwood, A. T. (1997) *Experiments in Ecology. Their Logical Design and Interpretation Using Analysis of Variance*. Cambridge: Cambridge University Press.

Valiela, I., McClelland, J., Hauxwell, J., Behr, P. J., Hersh, D. and Foreman, K. (1997) Macroalgal blooms in shallow estuaries: controls and ecophysiological and ecosystem consequences. *Limnology and Oceanography*, **42**: 1105–1118.

Viaroli, P., Azzoni, R., Bartoli, M., Giordani, G. and Taje, L. (2001) Evolution of the trophic conditions and dystrophic outbreaks in the Sacca di Goro lagoon (northern Adriatic Sea). pp. 467–475. In: *Mediterranean Ecosystems: Structures and Processes*. Faranda, F. M., Guglielmo, L. and Spezie, G. (eds). Milan: Springer-Verlag Italia.

21. Predicting Habitat Change in Poole Harbour Using Aerial Photography

Katie Born

Halcrow Group Ltd, Burderop Park, Swindon SN4 0QD

Predicting habitat change is essential to strategic coastal planning for estuaries to demonstrate compliance with the Habitat Regulations and biodiversity targets and to address its effects on hydrodynamics. Such predictions were made in the Poole Bay and Harbour Strategy Study. Historical trends in habitat change were estimated in Poole Harbour by interpreting georectified aerial photographs and mapping the extents of saltmarsh and reedbed using a GIS. Between 1947 and 1993, 245 ha (38%) of saltmarsh was lost and 47 ha (63%) of reedbed were gained. The rates of change varied across the harbour. Future habitat extents were estimated using continuous compounding rates of change and extrapolation based on the observed trends. The effects of historical and future sea level rise on the extent of saltmarsh were also estimated. The habitat change results are being included in a wider assessment of estuarine processes in the harbour to aid strategic coastal planning.

Introduction

Role of habitat change studies

Studying the extent of existing habitats in estuaries and predicting future habitat change is essential to strategic coastal planning by demonstrating compliance with the Habitat Regulations (the UK Conservation (Natural Habitats Etc.) Regulations 1994) and biodiversity targets. It also enables the potential effects of intertidal habitat change on flooding and navigation to be addressed because, for example, the conversion of saltmarsh to mudflat may affect the hydrodynamics of the estuary. The interlinking of habitat change, sea level rise and estuary form into strategies improves the basis for future flood defence planning and management of habitat resources.

The study described here was undertaken as part of a wider assessment of the estuarine processes considered in the Poole Bay and Harbour Strategy Study. Poole Harbour is a large estuary on the south coast of England, which is fed principally by the Rivers Frome, Piddle, Corfe and Sherford. It is almost an enclosed body of water and has a small tidal range (less than 2 m). Poole Harbour is designated as a Special Protection Area (SPA) and a Ramsar Site for its internationally important flora and fauna supported by its extensive tidal mudflats, saltmarshes, reedbeds, lagoons, freshwater marshes and wet grassland.

Saltmarsh

Saltmarsh vegetation develops in characteristic zones associated with saline or brackish tidal regimes (Burd, 1989; Rodwell, 2000). Poole Harbour contains over 20 different saltmarsh communities and sub-communities, including three Annex 1 habitats (listed in the European Habitats Directive 92/43/EEC) (Edwards, 2002).

Many saltmarshes are now dominated by common Cord Grass, *Spartina anglica*, which has spread naturally and by planting to aid sea defence and land-claim (Davidson, 1991; Raybould, 1997). It first started colonizing Poole Harbour in the 1890s and spread rapidly over a vacant niche on the mudflats (Gray *et al.,* 1991) to cover 800 ha by 1924 (Raybould, 1997). However, since the 1920s there have been substantial losses (Hubbard, 1965, cited by Bird, 1966; Raybould, 2000), which have also been observed in other swards in the south of England (Davidson, 1991). Potential causes of this decline may include:

- die-back of *S. anglica*
- invasion of other species (such as *Phragmites australis*)
- wave erosion
- sea level rise
- deliberate land reclamation
- other anthropogenic causes, such as pollution, dredging, etc. (Gray and Pearson, 1984; Gray *et al.*, 1991; Raybould, 1997).

Hubbard (1965, cited Gray and Pearson, 1984) estimated a loss of 170 ha from the 1920s to 1952, mainly from seaward bays, which may indicate that the harbour was affected by wave erosion. However, die-back was the principal cause of loss of *S. anglica* in Holes Bay (Hubbard, 1965, cited by Raybould, 1997). The changes in *S. anglica* extent has greatly affected the sedimentation process in Poole Harbour, first by accreting and consolidating sediment by rhizome growth during its gain, which deepened navigation channels, then by releasing it during die-back and erosion, which caused a shoaling of the channels (Raybould, 2000).

Reedbeds

Reedbeds are generally characterized by the overwhelming dominance of the common reed *Phragmites australis* (Rodwell, 1995; Cook, 2001), and form on predominantly wet or periodically flooded freshwater or tidal land. Some of the main reedbeds in Poole Harbour are found on either side of seawalls, which indicates that they have both freshwater and marine influences, although others have purely freshwater influences (Cook, 2001). Indeed, in Poole Harbour, *P. australis* has survived salinities of up to about 22 ppt, close to the experimentally determined limit (Rodwell, 1995).

Methodology

Measuring historic habitat extents from aerial photographs

Three sets of vertical aerial photographs of Poole Harbour were obtained from 1947, 1972 and 1993 (see Figure 1) covering the area shown in Figure 2. Following georectification from OS mapping, the images from each period were converted into a photo-mosaic and the extents of saltmarsh and reedbed were digitized on-screen using a geographical information system (GIS) software. Areas where there was uncertainty were verified by ground truthing, comparison with independent habitat maps prepared by Dorset Environmental Records Centre and, in the case of the older material, cross-comparison with recent, higher resolution colour photographs.

The total areas of saltmarsh and reedbed were calculated for the whole harbour and the six harbour sections using GIS, from which the past rates of change were derived.

Extrapolating from past change

A range of the future change in saltmarsh and reedbed up to 2053 was then predicted for the whole harbour study area and the six harbour sections based on the past trends. The extent of saltmarsh was estimated using logarithmic continuous compounding, because the magnitude of habitat loss depends on the initial area, gradually tending towards zero. The equations used were:

$$A = Pe^{in}$$
$$i = \log(A/P)/(n\log(e))$$

where P is the principal area, i is the rate of change, A is the area after *n* years, *n* is the number of years, e is the base of natural logarithms (2.71828... etc).

The extent of reedbed in 2053 was estimated using an arithmetic extrapolation of the annual rates of observed change. Continuous compounding rates of change were not used because reedbed gain is not necessarily associated with how much is present and compounding cannot predict changes in sections with no reedbed initially.

Although the suite of causal factors contributing towards saltmarsh loss was included in the extrapolations of future loss, it was possible to measure the area lost to deliberate land reclamation using GIS and to distinguish the area directly affected by sea level change.

Effects of sea level rise

The component of saltmarsh loss resulting directly from sea level rise was estimated independently. Maps of the saltmarsh distributions were overlaid on topographical data obtained from a 1998 LiDAR survey in GIS using a digital terrain model (DTM) of Poole Harbour. This enabled an estimate of the saltmarsh elevation and thus the

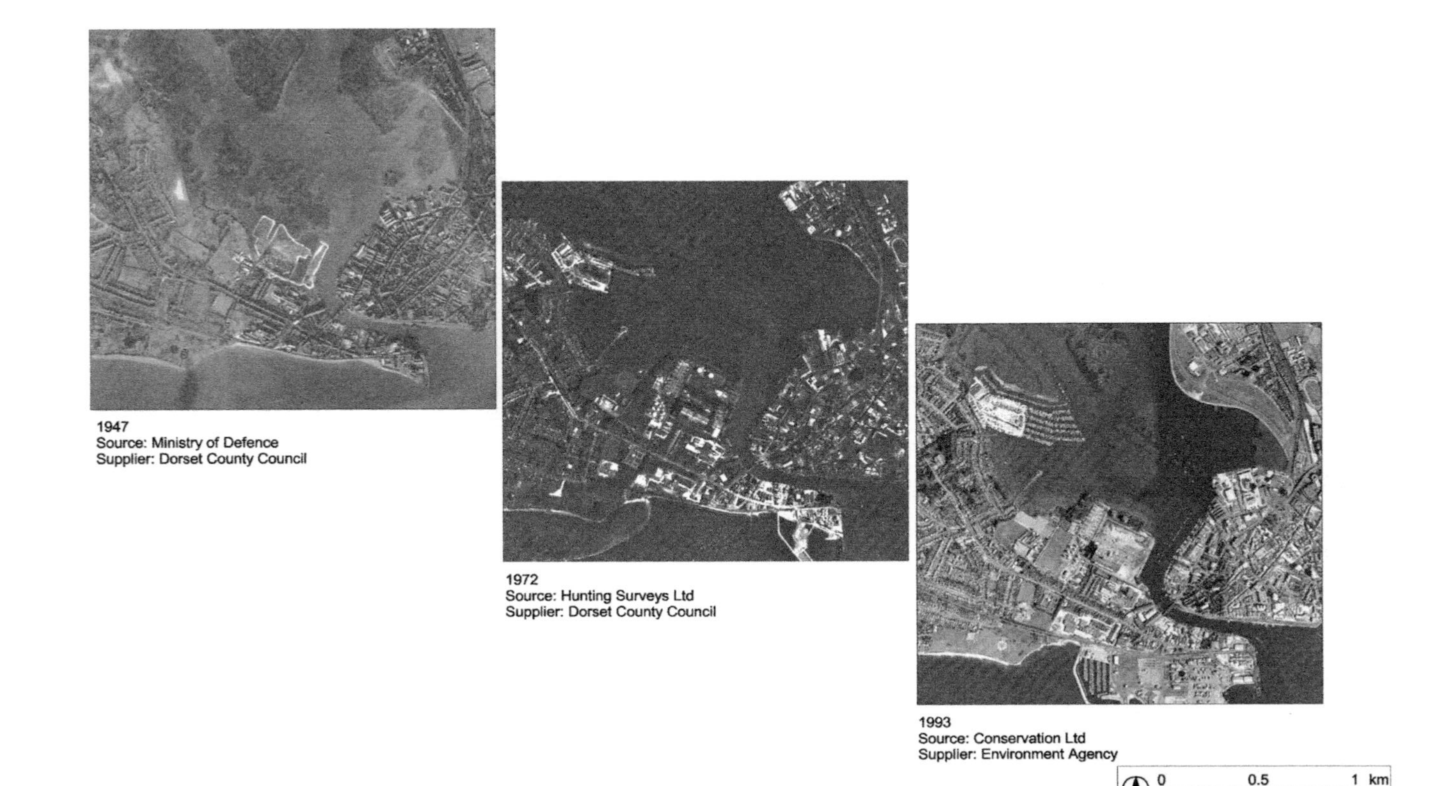

Figure 1 Aerial photography used for habitat mapping.

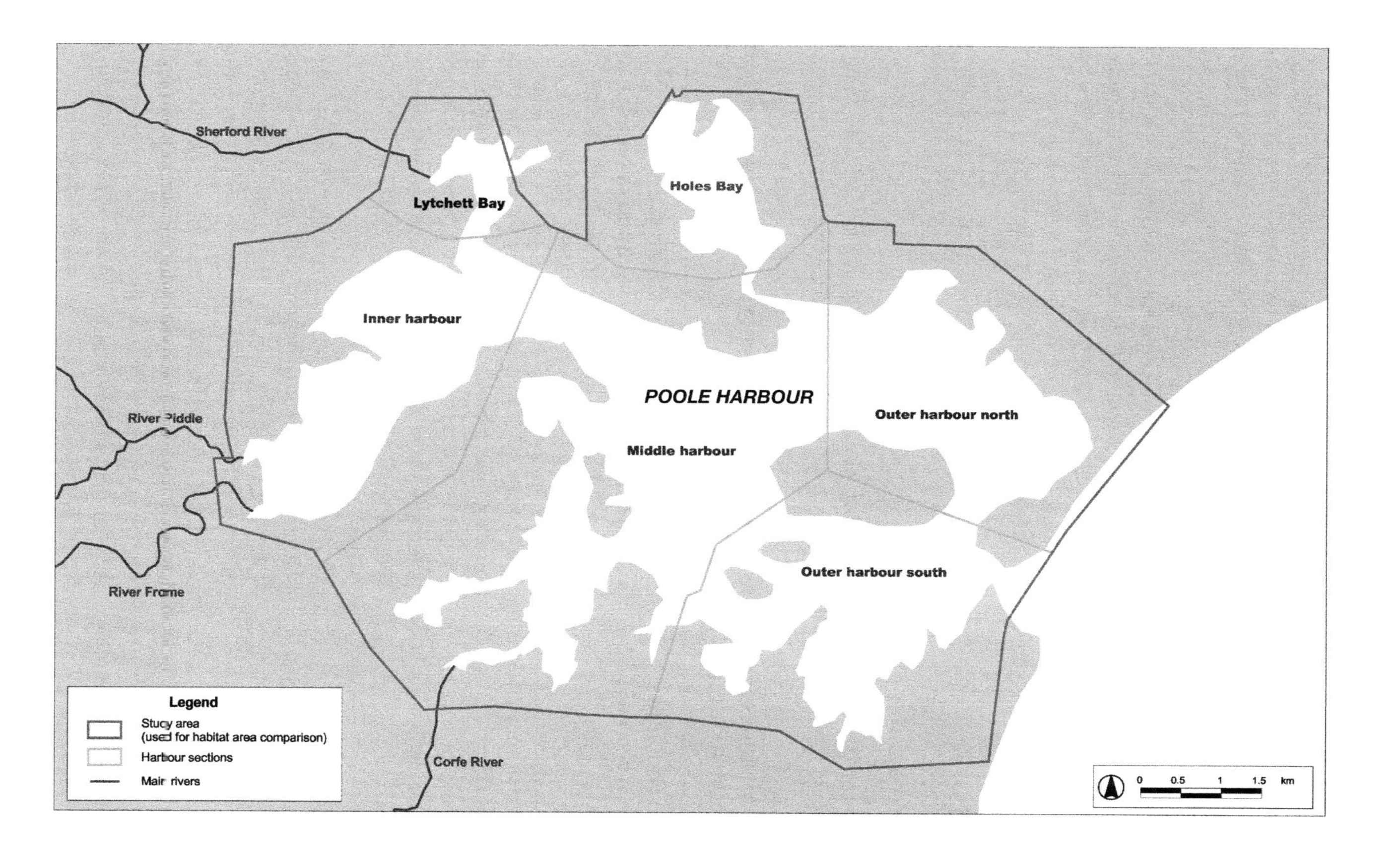

Figure 2 Study area.

percentage liable to be inundated for any given estimate of future sea level rise. Historic sea level rise was taken as a rate of 1.3 mm $year^{-1}$ and future sea level rise was taken as a rate of 5 mm $year^{-1}$ from 2003 (MAFF, 1999; website reference 1).

In estimating the effects of sea level rise, the lowest and highest extent of saltmarsh were taken to be the fifth and ninety-fifth percentile elevation values, to discount saltmarsh growing in marginal habitat and to remove the effects of any errors from the mapping or elevation data.

A range of saltmarsh areas affected by sea level change was calculated to allow for the differences in saltmarsh ability to migrate to higher elevations and the presence of a constraint to migration, such as a sea defence structure (Table 1). In other words, net loss is a balance between saltmarsh being submerged and lost from the lowest elevations and the extent of 'coastal squeeze' to the saltmarsh at the highest elevations.

Results

Observed habitat change

The extents of the saltmarsh and reedbed in the whole study area in 1947, 1972 and 1993 are shown on Figures 3 and 4. Between 1947 and 1993, the area of saltmarsh in the whole

Table 1 Methodology used to calculate areas of historic and future saltmarsh loss due to sea level change

Possible saltmarsh response to sea level change	Estimated historic saltmarsh loss due to sea level change (1947–93)	Estimated future saltmarsh loss due to sea level change (1993–2053)
No saltmarsh migration at a constrained or unconstrained site (erosion)	The lowest 0.060 m of saltmarsh, i.e. area between the lowest elevation (fifth percentile) and an elevation 0.060 m higher	The lowest 0.263 m of saltmarsh, i.e. area between the lowest elevation (fifth percentile) and an elevation 0.263 m higher
Saltmarsh migration at a constrained site (coastal squeeze)	The highest 0.060 m of saltmarsh, i.e. area between the highest elevation (ninety-fifth percentile) and an elevation 0.060 m lower	The highest 0.263 m of saltmarsh, i.e. area between the highest elevation (ninety-fifth percentile) and an elevation 0.263 m lower
Saltmarsh migration at an unconstrained site	Area of net saltmarsh loss would be nil	Area of net saltmarsh loss would be nil

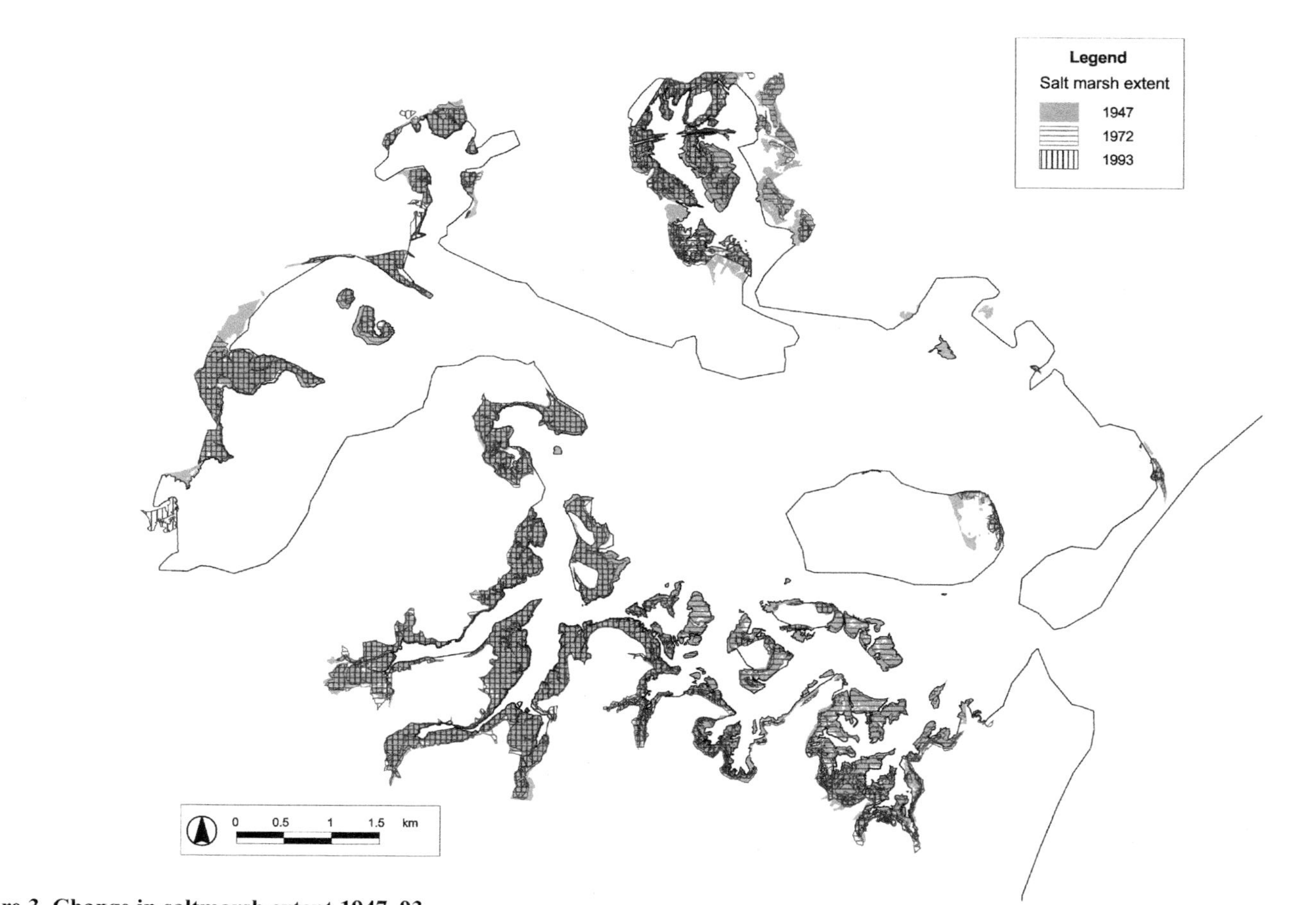

Figure 3 Change in saltmarsh extent 1947–93.

harbour study area decreased by 245 ha (38% of 1947 extent), whilst the area of reedbed increased by 47 ha (63% of 1947 extent) (Table 2). The rates of change varied over the years with saltmarsh loss accelerating and reedbed gain decelerating to virtually no change since 1972 (Figure 5).

Between 1947 and 1993, all six sections of the harbour experienced a loss in saltmarsh, although at different rates (Figure 6). The greatest rate of loss was in the Outer Harbour South (1972–1993) and the least was observed in Lytchett Bay (1947–1993). Likewise, the changes in reedbed also differed around the harbour over the years (Figure 6). Between 1947 and 1993, Inner Harbour, Middle Harbour and Outer Harbour North experienced a gain in reedbed and Lytchett Bay experienced a slight loss. However, between 1972 and 1993, only two sections experienced a gain in reedbed and two experienced a loss.

Extrapolation from past change

The results of extrapolating the 1947–93 and 1972–93 trends of habitat change to 2053 are shown in Table 4. Losses of saltmarsh are expected to be between 183 ha and 244 ha (47–63% of 1993 extent), whilst reedbed gains would be between 0 and 61 ha (0–50% of 1993 extent).

Effects of sea level rise

Between 1947 and 1993, up to 9 ha of saltmarsh were lost directly from sea level rise if migration did not occur, or up to 0.6 ha as a result of coastal squeeze (up to 3.5% and 0.2%, respectively of the observed saltmarsh loss). By 2053, the amount of saltmarsh estimated to be lost due to future sea level change would be up to 76 ha if migration does not occur, or up to 150 ha as a result of coastal squeeze. The study of saltmarsh elevations also found that between 1947 and 1993 most saltmarsh had been lost at the lowest range of elevations (Figure 7).

Summary of causes of habitat change

Several causes of habitat change were confirmed during the study (Table 4).

Table 2 Areas of saltmarsh and reedbed from historical aerial photographs in the whole harbour study area

Habitat	Observed area (ha)		
	1947	1972	1993
Saltmarsh	634	549	389
Reedbed	75	122	122
Total	709	671	511

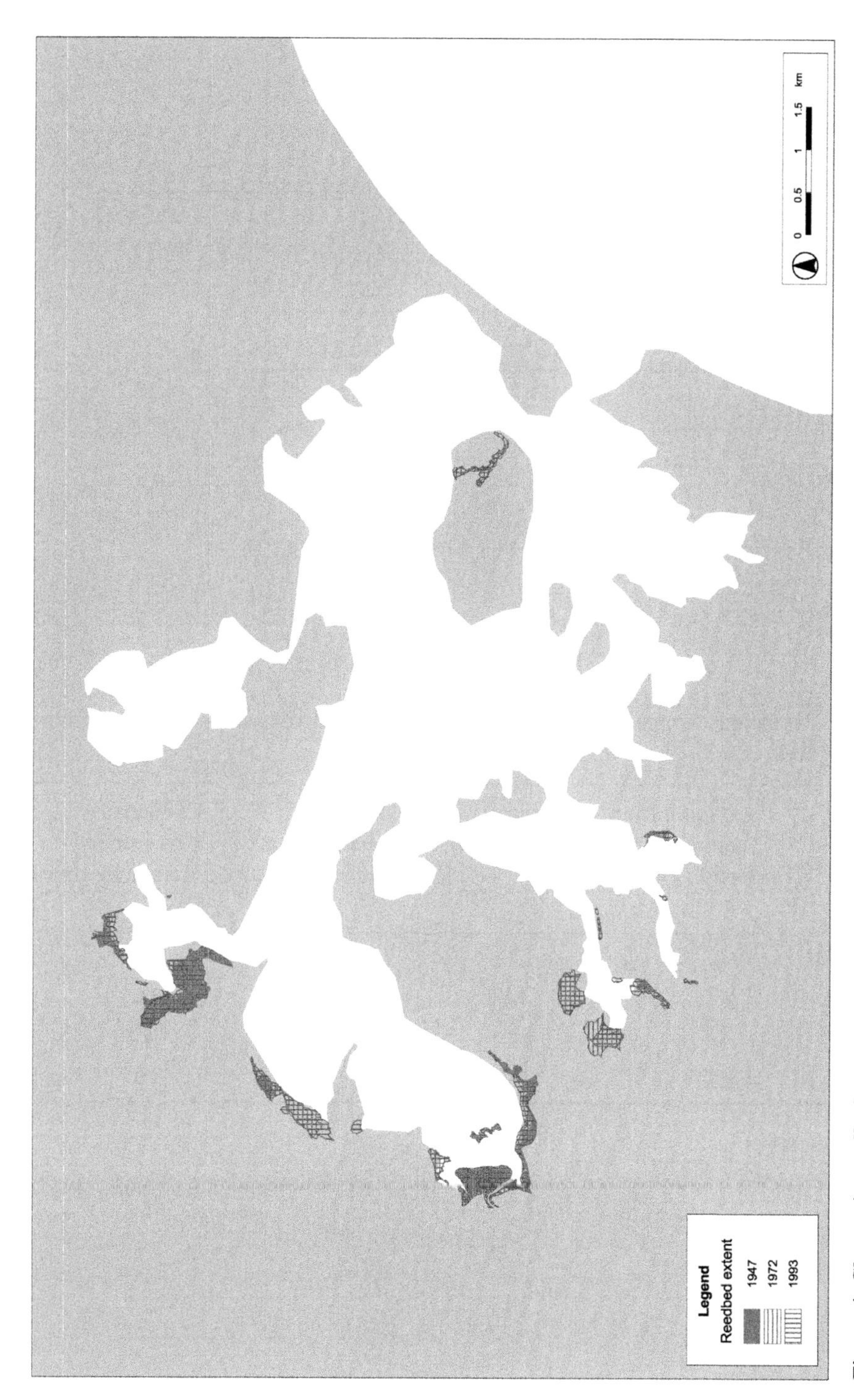

Figure 4 **Change in reedbed extent 1947 –93.**

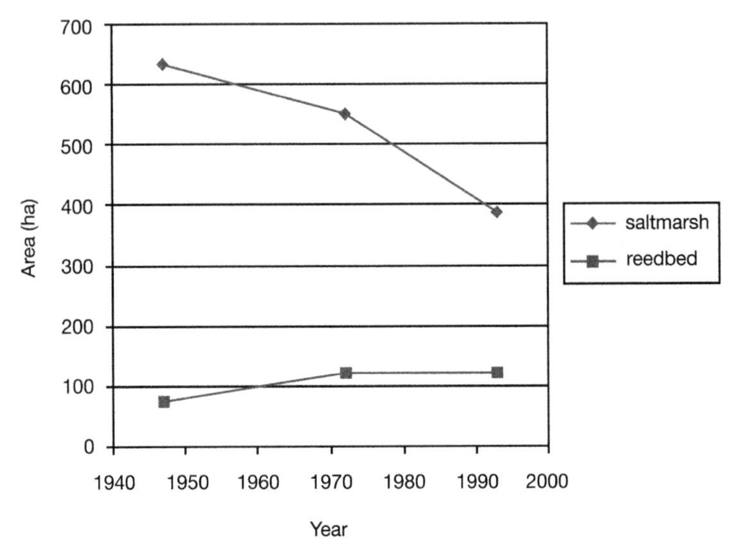

Figure 5 Observed habitat areas in the whole harbour study area.

Table 3 Predicted changes in saltmarsh and reedbed in the whole harbour study area

Habitat	Extrapolation from observed trend	Projected area in 2053 (ha)	Projected change in area since 1993 (ha)	Projected percentage change in area (%)
Saltmarsh	1972–93	145	-244	-63
	1947–93	205	-183	-47
Reedbed	1972–93	122	0.5	0.4
	1947–93	183	61	50

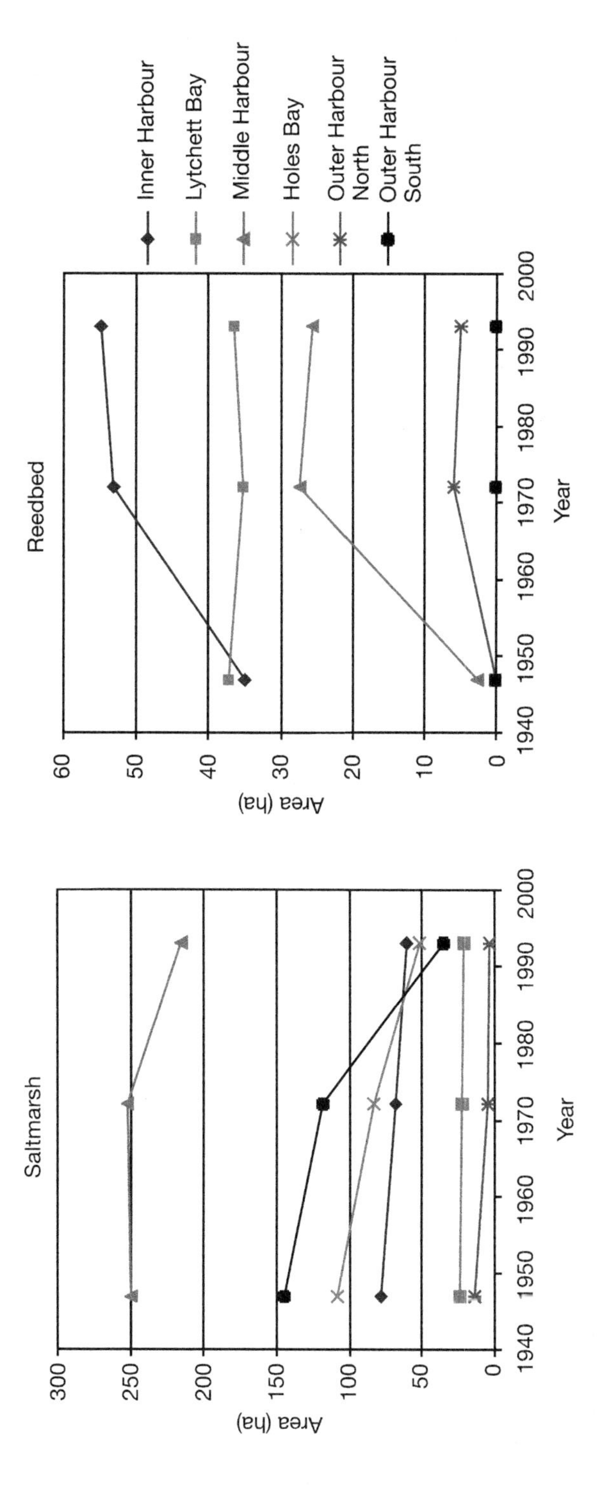

Figure 6 Area of observed saltmarsh and reedbed around Poole Harbour.

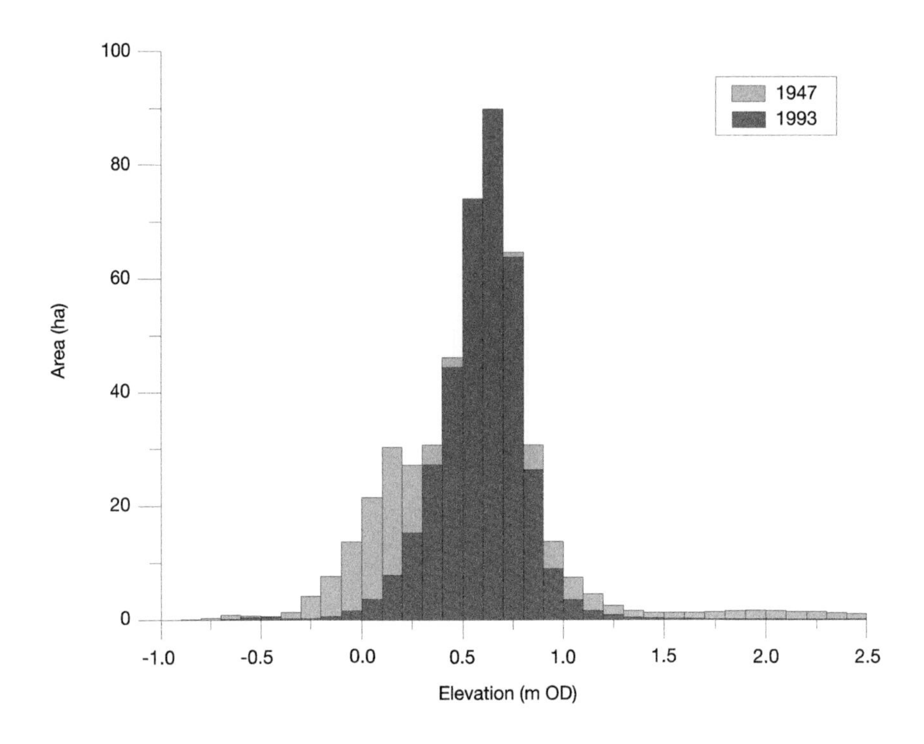

Figure 7 Distribution of saltmarsh with respect to elevation in 1947 and 1993.

Sensitivity analysis

Georectification of the aerial photographs and habitat mapping were estimated to have potentially introduced minimal error. However, sensitivity to scientist interpretation was considered to be the greatest potential cause of error, but the various verification methods described above reduced this.

Discussion

Review of methodology

The use of aerial photographs proved to be invaluable in measuring habitat change, as had been shown by Gray and Pearson (1984). However, several assumptions were made and there may have been opportunities for the introduction of error. The main assumption was that the observed rates of change from the past would continue unaltered in the future, but extrapolation of past change is not necessarily an accurate guide to the future. Also, extrapolation depends on which observed trend is used, and estimating the effects of sea level change depends on the response of the saltmarsh. Therefore, a range of results has been presented as a guide.

Table 4 Summary of the causes of habitat change

Potential cause of change in saltmarsh extent	Observed saltmarsh loss (1947–93) (ha)	Future saltmarsh loss (1993–2053) (ha)	Comment
Combination of causes (except land reclamation): • *Spartina* die-back • invasion by other species • wave erosion • sea level rise • other anthropogenic causes	245	183–244	This assumes that all of these processes will continue unchanged in the future Range of future loss given depends on which observed trends are used in the extrapolation
Spartina die-back	Note[1]	Note[2]	Evidence of this as most observed saltmarsh loss was experienced at the lowest elevations, where *Spartina* lies; maybe less important in future as most now lost
Invasion of saltmarsh by other species, e.g. *Phragmites*	Note[1]	Note[2]	Evidence of this as 22 ha of saltmarsh has converted to reedbed between 1947 and 1993
Wave erosion	Note[1]	Note[2]	Evidence of this as most decline observed in seaward parts of the harbour
Other anthropogenic causes, such as pollution, dredging, etc.	Note[1]	Note[2]	Direct evidence cannot be established using this methodology
Sea level rise	Up to 9 (erosion) or 0.6 (coastal squeeze)	Up to 76 (erosion) or 150 (coastal squeeze)	Evidence that 1947 saltmarsh was submerged by the 1993 sea levels. 1993 saltmarsh will be submerged by predicted 2053 sea levels. Range as it depends on the opportunity for saltmarsh migration. Likely to be dominating causal factor in the future and may result in more loss than predicted above
Land reclamation	30	Expected to be small	Removed from analysis of future change. Evidence of this in the past from aerial photographs and may occur in the future, but to a smaller extent due to current environmental regulations

[1]The amount of the saltmarsh loss directly caused by this individual factor is unknown.
[2]The amount of the future saltmarsh loss that will be directly caused by this individual factor is unknown.

Observed habitat change

This study has confirmed that several causal factors may have occurred in Poole Harbour in the last 50 years (see Table 4). The aerial photographs showed that 30 ha of saltmarsh were lost to deliberate land reclamation. However, an important reason for habitat changes in Poole Harbour is likely to be the decline of *Spartina anglica*, which has been observed by many other authors (see Bird, 1966; Gray and Pearson, 1984; Raybould, 2000). The histograms of saltmarsh elevations (Figure 7) show that most saltmarsh was lost at the lower elevations, which is the zone where *S. anglica* is mainly found (Burd, 1989). This may suggest that *Spartina* die-back and/or wave erosion occurred. Saltmarsh was invaded by other species, as 22 ha had converted to reedbed between 1947 and 1993.

A substantial reduction of saltmarsh occurred at seaward locations, which was also observed by Gray and Pearson (1984), which suggests that wave erosion has been a key cause of saltmarsh loss. Another likely cause is the direct effects of sea level change, causing up to 9 ha (4%) of the observed saltmarsh loss.

Predicted habitat change

In the future, the direct effects of sea level change are expected to be much greater than previously observed, as the rate of sea level rise is expected to be greater (MAFF, 1999). Saltmarsh loss directly from sea level rise is expected to be up to 150 ha by 2053, and it is likely to become the dominating causal factor for saltmarsh loss in the future and may mean that the total saltmarsh loss may be greater than that made by the extrapolations. The impact of sea level rise depends on the saltmarsh ability to migrate to higher elevations as fast as sea level rise. In the future, the impact of 'coastal squeeze', where constraints would stop saltmarsh from retreating, will be greater than that of submergence at the lower elevations as most saltmarsh communities now lie at the higher elevations.

Habitat resource planning

The observed and predicted changes to these internationally important habitats have implications for habitat resource planning. To comply with the Habitats Regulations and Biodiversity Action Plan targets, provision should be made for recreating saltmarsh to replace that which is lost. This study could be used to identify areas where further monitoring of habitat change, replanting measures or managed realignment of coastal defences might be appropriate.

Conclusions

This is believed to be the most reliable and up to date study of saltmarsh loss and reedbed gain in Poole Harbour, although the results should be considered as a guide to possible future change. It has also confirmed the several causes of past habitat change and that sea level rise is likely to become a dominating cause of saltmarsh loss in the future.

The study has implications for sustainable estuary management since the replacement of the lost habitat will be required to comply with nature conservation obligations, such as through the managed realignment of coastal defences.

Acknowledgements

I would like to thank Laurence Banyard, Robert Harvey and Richard Westaway at Halcrow Group Ltd for their work on the study, and also Bryan Edwards and Carolyn Steele (Dorset Environmental Records Centre), Steve Spring (Dorset County Council), the Environment Agency and Sue Barton (English Nature) for providing information.

References

Bird, E. C. F. (1966) *Physiographic Changes and Land Reclamation in Poole Harbour.* Melbourne: Department of Geography, University of Melbourne.

Burd, F. (1989) *The Saltmarsh Survey of Great Britain. An Inventory of British Communities.* Peterborough: Joint Nature Conservancy Committee.

Cook, K. (2001) *Poole Harbour Reedbeds – Year 2000 Survey Summary Report.* Purbeck Reedbed Working Group.

Davidson, N. C. (ed.) (1991) *Nature Conservation and Estuaries in Great Britain.* Peterborough: Nature Conservancy Council.

Edwards, B. (2002) The Vegetation and Flora of Poole Harbour. Draft report for the Poole Harbour Study Group. Dorchester: Dorset Environmental Records Centre.

Gray, A. J., Marshall, D. F. and Raybould, A. F. (1991) A century of evolution in *Spartina anglica. Advances in Ecological Research,* **21**: 1–62.

Gray, A. J. and Pearson, J. M. (1984) *Spartina* marshes in Poole Harbour, Dorset, with particular reference to Holes Bay. pp. 11–14. In: *Spartina anglica in Great Britain.* Doody, P. J. (ed.). *Focus on Nature Conservation,* No.5. Huntingdon: Nature Conservancy Council.

MAFF (1999) *Project Appraisal Guidance 3, Economic Appraisal.* London: Ministry for Agriculture, Fisheries and Food.

Raybould, A. F. (1997) The history and ecology of *Spartina anglica* in Poole Harbour. *Dorset Proceedings,* **119**: 147–158.

Raybould, A. F. (2000) Hydrological, ecological and evolutionary changes associated with *Spartina anglica* in Poole Harbour. In: *British Saltmarshes.* Sherwood, B. R., Gardiner, B. G. and Harris, T. (eds). Cardigan: Forest Text.

Rodwell, J. S. (ed.) (1995) *British Plant Communities* Volume 4: *Aquatic Communities, Swamps and Tall-herd Fens.* Cambridge: Cambridge University Press.

Rodwell, J. S. (ed.) (2000) *British Plant Communities* Volume 5: *Aquatic Maritime Communities and Vegetation of Open Habitats.* Cambridge: Cambridge University Press.

Website reference:

www.pol.ac.uk/psml/datainfo/rlr.trends permanent service for mean sea level, hosted by Proudman Oceanographic Laboratory, based on 2001 data.

22. Poole Harbour European Marine Site

Helen Powell

English Nature Dorset Team, Slepe Farm, Arne, Wareham, Dorset BH20 5BN

The features of interest that make Poole Harbour internationally important are explained and the significance of Poole Harbour in its context as part of a network of special sites across Europe is discussed. The Habitats Regulations make special provisions for European Marine Sites and the requirements of Regulations 33 and 34 in particular are explained.

Natura 2000

The term Natura 2000 comes from the 1992 EC Habitats Directive[i] and is the title for a network of legally protected areas across the European Community. The sites that make up the Natura 2000 network have been designated to conserve natural habitats and species of wildlife that are rare, threatened or vulnerable across the Atlantic Bio-geographic Region. The Natura 2000 network represents the very best nature conservation sites across Europe.

The Natura 2000 network is made up of two types of site which originate from two different but complementary European Directives. The Birds Directive[ii] requires member states to designate Special Protection Areas (SPAs) where an area supports significant numbers of wild birds and their habitats. The Habitats Directive requires member states to designate Special Areas of Conservation (SACs) where a site supports outstanding examples of habitats that are characteristic of the Atlantic Bio-geographic Region, or if it supports rare, endangered or vulnerable species of plants or animals. Where a SPA or SAC includes any part of the sea or seashore, it is also referred to as a European Marine Site.

Because of the immense productivity of estuaries generally, combined with the UK's relatively mild climate and position on the western edge of the European landmass, estuaries are extremely important wintering areas and stopping points for many migratory waterbirds. The UK has a significant contribution to make to the international conservation of populations of wild birds and the habitats on which they depend, and this has been recognized in the designation of many estuarine and coastal SPAs. The UK's SPA network supports an average of over 2 million non-breeding waterbirds (Stroud *et al.*, 2001) and contributes to the

[i] Council Directive 92/43/EEC on the conservation of natural habitats and wild flora and fauna.
[ii] Council Directive 79/409/EEC on the conservation of wild birds.

suite of sites across Europe. Those SPAs and SACs that include a marine element are represented in the UK by the 39 European Marine Sites which amount to over 0.5 million ha.

Poole Harbour

The intertidal mudflats, sandflats and marshes of Poole Harbour support large numbers of wintering wildfowl and waders that are of both national and international significance. It was notified as a Site of Special Scientific Interest (SSSI) in 1990 and was classified as a SPA in 1999. As the SPA includes the intertidal area, Poole Harbour is also a European Marine Site. The extent of the Poole Harbour European Marine Site is just over 1983 ha and lies between Mean Low Water and Highest Astronomical Tides.

Poole Harbour qualifies as a SPA by supporting:

- **internationally important populations of regularly occurring Annex 1 species**[iii]
  the site supports internationally important populations of Avocet *Recurvirostra avosetta*, Mediterranean Gull *Larus melanocephalus* and Common Tern *Sterna hirundo*
- **internationally important populations of regularly occurring migratory species**
  the site supports internationally important numbers of regularly occurring migratory Black-tailed Godwit *Limosa limosa* and Shelduck *Tadorna tadorna*
- **an internationally important assemblage of waterfowl**
  Poole Harbour regularly supports over 20,000 birds (Cranswick *et al.*, 1999).

The intent and specific requirements of the Birds Directive and the Habitats Directive are transposed into UK legislation by the Habitats Regulations[iv]. The Habitats Regulations form the basis for establishing, protecting and managing SPAs and SACs in the UK. In terms of European Marine Sites, Regulations 33 and 34 are key, because they make special provisions for European Marine Sites.

Regulation 33

Regulation 33 requires that as soon as possible after a site becomes a European Marine Site, English Nature must advise other relevant authorities on the specific conservation objectives for the site and identify any operations which may cause deterioration of natural habitats or disturbance of species for which the site has been designated.

English Nature set out the Regulation 33 advice for Poole Harbour European Marine Site in November 2000. An inherent part of this advice is the favourable condition table. This

---

[iii] Species listed in Annex 1 of the Birds Directive are in danger of extinction, rare or vulnerable and are the subject of special conservation measures concerning their habitat.

[iv] The Conservation (Natural Habitats &c.) Regulations 1994.

identifies attributes of the site, against which the impact of development proposals and other plans or projects may be assessed. It also provides a framework for English Nature to report on the condition of Poole Harbour SPA to the European Commission. In the case of many terrestrial European Sites, sufficient is known about the preferred or target condition of qualifying habitats to be able to define measures and associated targets in condition monitoring. However, with European Marine Sites, less is known about habitat condition and in many cases, and certainly in the case of Poole Harbour European Marine Site, existing condition needs to be established through baseline survey.

Food availability is one of the attributes that requires monitoring by English Nature to confirm that the interest features of Poole Harbour SPA are in favourable condition. In Autumn 2002, the Centre for Ecology and Hydrology and EMU Ltd, respectively, were commissioned to undertake a baseline survey and subsequent analysis of bird food availability. The objectives of the project were to establish the existing bird prey invertebrate abundance and biomass in the intertidal sediment communities of Poole Harbour. This baseline information will be used to compare future surveys, so that any significant changes in prey availability may be detected, bird condition and mortality over a winter season predicted and and may also identify whether certain bird species are at risk from insufficient food.

It is a case in point of how we need to ensure that science informs the management of our most special sites. With the Habitats Regulations requiring a precautionary approach with respect to SPAs and SACs, the ability to be more predictive may allow improvements to be targeted more efficiently and potential threats to the site to be identified at an early stage. The Poole Harbour Study Group has so far been an enormously valuable partnership, in that the organizations and individuals represented on the Study Group are working together to understand more about Poole Harbour and its condition. Examples of where the Study Group has been instrumental in furthering our understanding of the site are in the publication of the *Poole Harbour Flora*, the *Reedbed Assessment*, and the soon to be published, *Wader Roost Survey*.

Regulation 34

Regulation 34 is another key aspect of the Habitats Regulations in terms of European Marine Sites. Regulation 34 provides for the establishment of an agreed management scheme for a European Marine Site. The intention of a management scheme is to provide a mechanism for resolving management issues and to set a framework in which activities that occur within a site are managed either voluntarily or through regulation, in order to achieve the conservation objectives of the European Marine Site.

Whilst the requirement to provide advice under Regulation 33 is a statutory requirement, the establishment of a management scheme is at the discretion of the relevant consenting authorities with responsibility for the management of the site. On the great majority of sites, the development of a management scheme is a sensible and practical tool in the management of the site. Indeed, in the broadest sense, on every European Marine Site, there is likely to be some form of management scheme, or the revision of an existing plan; the scale will depend on the number and complexity of potential conflicts and management issues.

European LIFE funding established the UK Marine SACs Project, which amongst other things piloted the production of management schemes for 12 marine SACs, selected as being representative of a range of management issues encountered on the national series of 39 European Marine Sites. In Poole Harbour, there is an existing Aquatic Management Plan which was first published in 1992. The Poole Harbour Steering Group is about to embark on a revision of this plan. It is anticipated that where a particular management approach has worked well, this should be taken forward into the revision of the plan. The Steering Group is in a strong position now, as we can reflect and learn from the successes and difficulties encountered, not just with our own existing management plan for Poole Harbour, but also from the other management schemes piloted for European Marine Sites elsewhere. It is undoubtedly an opportunity to build on existing partnerships, raise awareness of the importance and value of Poole Harbour, and ultimately, to help to achieve and sustain the favourable condition of this internationally important site.

References

Cranswick, P., Pollitt, M., Musgrove, A. and Hughes, B. (1999) *The Wetland Bird Survey 1997–98 Wildfowl and Wader Counts*. British Trust for Ornithology, WWT, RSPB and JNCC.

Stroud, D. A., Chambers, D., Cook, S., Buxton, N., Fraser, B., Clement, P., Lewis, P., McLean, I., Bake,r H. and Whitehead, S. (2001) *The UK SPA Network: Its Scope and Content.* Volume 1. *Rationale for the Selection of Sites*. Peterborough: Joint Nature Conservation Committee.

Conclusion: Science, Development and Management

John Humphreys[1] and Vincent May[2]

[1]University of Greenwich, Old Royal Naval College, Greenwich, London SE10 9LS

[2]School of Conservation Sciences, Bournemouth University, Talbot Campus, Fern Barrow, Poole, Dorset BH12 5BB

In this book, the contributors have demonstrated two general features of Poole Harbour: first, its considerable value as a significant and novel natural coastal environment with an important role in terms of biodiversity; second, the range and intensity of anthropogenic impacts on the harbour as a consequence of burgeoning development-related pressures ranging from tourism and leisure through industry and port activities to waste disposal and even hydrocarbon extraction. In looking to the future we must acknowledge that these twin features of the harbour, while they generate considerable potential conflict, cannot be seen as distinct from one another in terms of harbour ecology or management solutions. Rather environment and development as the Brundtland Commission observed (World Commission on Environment and Development, 1987) are inexorably linked and interrelated. Poole Harbour demonstrates this link not to be a purely recent phenomenon. Archaeological research supported by Poole Maritime Trust and the Poole Harbour Heritage Project shows the modern nature of the harbour to have been shaped by human communities over several millennia through land reclamation, port construction and modifications to land use and landscape within the harbour's catchment. However, the rate and intensity of anthropogenic alteration has accelerated during the past two centuries.

While development cannot flourish on an unsustainable environmental resource base, it is nevertheless inevitable, and is indeed an important contributor to human benefit on local, regional and international levels. Resolving the conflicts between environment and development cannot, therefore, be achieved by unco-ordinated institutions and policies, and it is in this context that the concept of integrated coastal zone management (ICZM) has emerged (World Bank, 1993; European Science Foundation, 2002). Cicin-Sain (1993) emphasized that 'integration' in coastal management is essential at several levels, one being integration among disciplines. Sustainable ICZM depends on understanding the social-economic-cultural-legal processes which bring about change, produce impacts and recognize opportunities for sustained development. Poole Harbour is managed by the Poole Harbour Commissioners whose "statutory duty is to conserve, regulate and improve" the harbour. Their Environmental Policy Statement emphasizes that they recognize "the special position of Poole Harbour as a natural asset and will continue to promote its sustainable use, balancing the demands of its natural resources and resolving conflicts of interest". An Aquatic Management Plan designed to zone activities throughout the harbour is already in effect and managed through a committee which brings together the Commissioners, the local authorities and bodies such as English Nature to monitor and resolve such conflicts as they arise. The land around the harbour and its catchment where many activities occur which can profoundly affect the ecology of the harbour, are subject to both planning legislation and also to a wide range of land designations which derive from both national and European regulations. Some of these specify the data which must be collected to assure the proper implementation of the regulations and this in turn provides a role for scientific research.

Sustainable management cannot in itself be effective unless established on a scientific foundation. Much scientific work within the harbour has been undertaken in order to understand issues such as water quality and shoreline change, which have implications for the sustained ecological and economic well-being of the harbour. However, because these studies have often been focused upon specific (and often pressing) problems of public concern, such as planning issues, or are focused primarily on monitoring in relation to statutory requirements, they stand apart from other research being carried out within the harbour. In this context, in order to make progress in terms of the science of the harbour, a more coherent scientific effort should combine a whole system focus with a multidisciplinary approach. Such an approach would improve our understanding of the Poole Harbour ecosystem as a whole in terms of its geomorphology, hydrology and physico-chemical characteristics and their relationship to biodiversity, species distribution, population dynamics and community ecology. Integral to such work would be an examination of the environmental impacts and interactions of anthropogenic agents on the ecosystem, and an enhanced understanding of how the human community responds to an improved understanding of the dynamics of the harbour system.

We hope that this book demonstrates that much has already been achieved in relation to such strategic themes. But much also remains to be done. In particular, while over the last few years important projects have considerably enhanced our knowledge of the components of the system, there remains the need for a deeper understanding of the harbour's ecology in the sense of establishing some of the key causal interrelations across physical, chemical and biological phenomena (anthropogenic and natural). If we can gain a greater grasp of causality in the Poole Harbour system, then we can begin to understand the consequences of anthropogenic agents and interventions in ways which would provide a stronger basis for decision-making and conflict resolution.

We recognize that Poole Harbour research, if set only in its local context, is unlikely to draw considerable and sustained additional research resources much beyond those already being invested. However, we believe that Poole Harbour has the potential to become more prominent in national and international research terms. The number of agencies currently committed to the harbour is not insignificant and neither is the aggregate value of their commitment. This book provides ample evidence of substantial resources of money and time invested in Poole Harbour research by organizations such as English Nature, the Environment Agency, Poole Harbour Commissioners, Wessex Water, Natural Environment Research Council among others – and through the work of academics and students at the Universities of Bournemouth, Greenwich, Southampton and Swansea. We suggest that, notwithstanding the particular statutory or other interests represented by such organizations, improved co-ordination of such work sometimes in the form of multi-agency projects, would lead to better returns in relation to our overall understanding of the harbour, whilst also of necessity continuing to meet the particular needs of the individuals and agencies involved.

Such co-ordination, in so far as it results in the improved elucidation of scientific and management issues may also provide a basis for drawing new resources from national or European sources for scientific research. To achieve this potential, Poole Harbour must be situated in its international context. In the Introduction, we noted that immediately adjacent to Poole Harbour is some of the most expensive real estate in the world. The fact that this real estate lies in such close proximity to internationally significant bird populations gives force to our assertion that Poole Harbour represents *par excellence* the environment-development debate in microcosm. It is not only these features, however, that provide Poole Harbour with international significance. The harbour has an established history in terms of the naturalization of non-indigenous species – indeed five of the chapters in this book focus on such invasions. Moreover, it has been asserted (English Nature, 1994) that Poole Harbour has the highest summer temperature maxima of any significant size marine water mass in the UK. Currently, in marine science, ecological community changes such as the naturalization of tropical species in temperate waters are being attributed to climate change. Arguably, the Manila Clam population in Poole Harbour is a case in point. In any event, lack of systematic knowledge of ecosystem responses to climatic change is widely recognized as an important scientific and indeed political issue (European Science Foundation, 2002). On the basis of existing Poole Harbour literature and historical data combined with improved co-ordination of effort, Poole Harbour could provide a site of considerable utility as a case study in terms of climate change related ecological work.

In any event, we hope and expect that the Poole Harbour Study Group will continue to encourage research on Poole Harbour and moreover facilitate the co-ordination of that research, and that this book, by drawing together reports on a wide range of aspects of the ecology of the harbour so as to make accessible a better overview than would otherwise be available, will further assist with that effort.

References

Cicin-Sain, B. (1993) Sustainable development and integrated coastal zone management. *Ocean and Coastal Management*, **21**: 11–44.

English Nature (1994) *Important Areas for Marine Wildlife around England*. Peterborough: English Nature.

European Science Foundation (2002) *Integrating Marine Science in Europe*. Strasbourg: European Science Foundation.

World Bank (1993) *Noordwijk Guidelines for Integrated Coastal Zone Management.* Washington DC: World Bank Environment Department, Land Water and Natural Habitats Division.

World Commission on Environment and Development (1987) *Our Common Future.* Oxford: Oxford University Press.

Index

Printed and bound by CPI Group (UK) Ltd, Croydon, CR0 4YY
08/05/2025
01864933-0003